Micromachined Mirrors

THE KLUWER INTERNATIONAL SERIES IN MICROSYSTEMS

Consulting Editor: Stephen Senturia
Massachusetts Institute of Technology

Volumes published in
MICROSYSTEMS

Heat Convection in Micro Ducts
Yitshak Zohar
ISBN: 1-4020-7256-2

Materials & Process Integration for MEMS
Francis E.H. Tay
ISBN 1-4020-7175-2

Microfluidics and BioMEMS Applications
Francis E.H. Tay
ISBN: 1-4020-7237-6

Optical Microscanners and Microspectrometers Using Thermal Bimorph Actuators
Gerhard Lammel, Sandra Schweizer, Philippe Renaud
ISBN 0-7923-7655-2,

Scanning Probe Lithography
Hyongsok T. Soh, Kathryn Wilder Guarini, Calvin F. Quate
ISBN 0-7923-7361-8

Microsystem Design
Stephen Senturia
ISBN: 0-7923-7246-8

Microfabrication in Tissue Engineering and Bioartificial Organs
Sangeeta Bhatia
ISBN 0-7923-8566-7

Microscale Heat Conduction in Integrated Circuits and Their Constituent Films
Y. Sungtaek Ju, Kenneth E. Goodson
ISBN 0-7923-8591-8

Micromachined Ultrasound-Based Proximity Sensors
Mark R. Hornung, Oliver Brand
ISBN 0-7923-8508-X

Bringing Scanning Probe Microscopy Up to Speed
Stephen C. Minne, Scott R. Manalis, Calvin F. Quate
ISBN 0-7923-8466-

Microcantilevers for Atomic Force Microscope Data Storage
Benjamin W. Chui
ISBN 0-7923-8358-3

Micromachined Mirrors

by
Robert Conant, Ph.D.

Distributors for North, Central and South America:
Kluwer Academic Publishers
101 Philip Drive
Assinippi Park
Norwell, Massachusetts 02061 USA
Telephone (781) 871-6600
Fax (781) 681-9045
E-Mail: kluwer@wkap.com

Distributors for all other countries:
Kluwer Academic Publishers Group
Post Office Box 322
3300 AH Dordrecht, THE NETHERLANDS
Telephone 31 786 576 000   e-ISBN
Fax 31 786 576 254
E-Mail: services@wkap.nl

  Electronic Services < http://www.wkap.nl>

Conant, Robert (Robert A.), 1970-
    Micromachined mirrors/by Robert Conant.
        p.cm.--(The Kluwer international series in microsystems)
    ISBN 978-1-4419-5325-4
        1.Micromachining. 2.Mirrors. 3.Optical instruments--Design and construction.
    I. Title. II. Series.

    TJ1191.5.C665 2003
    681'.428--dc21

                                                        2002035706

Printed on acid-free paper.

Printed in the United States of America.

Contents

Micromachined Mirrors

by
Robert Conant, Ph.D.

Acknowledgements

I would like to thank the many people who have contributed to this work. First and foremost, thank you to Professor Richard S. Muller, who was an extraordinary mentor during my graduate career. He warmly welcomed me to the University of California and has since consistently impressed me with his dedication to his students and his pursuit of scientific understanding. Professor Muller contributed greatly to this work – both in the understanding of scanning mirrors and in the writing. I am forever indebted to him for his support. He has been a role model for me, not only professionally, but personally.

This work was the result of collaboration with many excellent researchers at the University of California, Berkeley and Davis.

Professor Kam Y. Lau guided me and the other members of our research group towards engineering excellence. His inspiration in science, engineering, and business has directed my work towards real-world applications of cutting-edge engineering. Professor Lau's sharp and creative mind continues to inspire me to aspire to the highest standards.

Jocelyn Nee and Matthew Hart were both instrumental in the formulation and execution of this work. Jocelyn was closely involved with the Tensile Optical Surface work described in Chapter 4 and has been helpful in discussions about mirror design and fabrication. Matthew Hart conceived, built, and perfected the Twyman-Green interferometer that enabled dynamic characterization of moving microstructures that has shifted the focus of micromirror design from pretty pictures to high-quality components.

Professor Olav Solgaard, Paul Hagelin, and Uma Krishnamoorthy contributed to my understanding of surface micromachining (described in Chapter 3), and were wonderful collaborators on the raster-scanning projection video display described in Chapter 5.

xvi

Peter Jones provided insights into the mechanical descriptions of mirror dynamic deformation that were helpful in the formulations of Chapter 2.

Many thanks to those who did not contribute directly to this work, but who have helped my understanding of MEMS and micromachining processes: Professor Kristofer S.J. Pister, Michael Helmbrecht, Chris Keller, Michael Cohn, Roger Hipwell, Behrang Behin, Joe Seeger, Michel Maharbiz, Satinderpall Pannu, Carl Chang, Patrick Riehl, Veljko Milanovic, Dave Horsley, Jim Bustillo, Bob Hamilton, and Katalin Voros.

Finally, I would like to give my most sincere thanks to my family for providing continued support, encouragement, and love throughout my life. I owe to you all my deep commitment to education and my ability to achieve my dreams. Dale, Martha, Rich, Pat, Beth, and Brittany, thank you.

Chapter 1

Introduction

Optical systems have, over the past 50 years, revolutionized the way we interact with information, and now have a wide range of uses in telecommunications, information display, and metrology. The advances in optical science at the beginning of the 20th century laid the foundation for the understanding of optical particle and wave properties, and in the 1950's the invention of the laser allowed, for the first time, practical commercial and industrial applications of coherent optical systems. Coupled with the advent of high-speed electronics that resulted from the invention of the transistor in the 1950's and the need for high-speed communication resulting from rapid adoption of the high-volume, high-performance, low-cost silicon planar electronics manufacturing technologies, optical technologies began to revolutionize telecommunications and information transmittal in the 1980's. Today fiber-optic systems can routinely transmit information at rates of more than a terahertz.

Through this same period there was another line of progression of information transmittal using optical systems: transmittal of information to people via their highest-bandwidth sensor, the eye. These technologies included a method of using light to store images on paper (the photograph) and using light to create images from electrical signals (the television).

Today optical systems are used in a remarkable range of applications, many of which use light to translate information from one form to another. Everyday applications of optical systems are abundant, including optical systems that use light to translate electrical signals to visible images (video displays, laser printers), and optical systems that use light to translate from visible images to electrical signals (digital cameras, barcode scanners). Common industrial applications of optical systems use light to translate from electrical signals to mechanical structures (photolithography, stereo lithography), and from mechanical structures to electrical signals (interferometers and other optical metrology systems, CD-ROM and DVD optical data storage). Chemical, biological, and geological applications of optical systems are also common.

Miniaturization of optical components has enabled many of these applications. Lasers have gone from expensive table-top systems to diode lasers a fraction of a millimeter on each side and optical sensors are now routinely made with dimensions of only a few square micrometers. Integrated-circuit manufacturing techniques pioneered for the manufacture of high-speed low-cost electronics has enabled many of these advances in optical systems. Many of the other optical components used for common optical systems, however, have not seen the same increase in performance and decrease in cost associated with miniaturization.

1.   OPTICAL SCANNERS

One such optical component that has not seen the benefits of miniaturization is the optical scanner, which achieves the function of creating a time-varying angular deflection of light. Numerous solid-state methods of optical scanning have been used (including acoutso-optic and electro-optic) to achieve the benefits of miniaturization, but a large fraction of the optical scanners in use today, in commercial, scientific, and industrial applications, utilize mechanical rotating mirrors for optical beam deflection because they are capable of high-resolution, large-angle scanning with little wavelength sensitivity.

The critical performance specifications of optical scanners include optical properties (reflectivity, mirror flatness, scan angle, and optical resolution), cost (ease of manufacture, material requirements), power consumption, and speed. Uses of optical scanners can be divided into two broad categories: resonant scanning (for high-speed cyclic scanning), and steady-state beam steering (for redirecting an optical beam to a particular angle).

Scanning mirrors are used in a number of applications, including:
– Laser printers. Rotating polygonal mirrors scan a modulated beam across an electrostatic drum that is used to pick up toner and place it on paper. These scanners are typically five-sided and rotate at 30,000 rpm, giving a linear beam scan at 2500 Hz.
– Projection video displays. Two mirrors are scanned in orthogonal directions to create a raster scan, and red, green, and blue modulated light beams are reflected onto a screen or projected directly into the retina in a raster-scanned pattern to generate a video image. These scanners scan at frequencies up to 29 kHz.

- Scientific instruments. Albert Michelson used a scanning mirror to measure the speed of light in the 1920's, and scanning mirrors are still often used in scientific instruments. Macro-scale galvanometric scanners are commercially available at scan speeds up to 2 kHz with large scan angles; +- 45 degrees optical scan angle is not uncommon.

2. MICROELECTROMECHANICAL SYSTEMS (MEMS)

Conventional mechanical scanners, although quite useful for many applications, have significant performance limitations due to their size. Microelectromechanical systems (MEMS) technology – a set of manufacturing techniques broadly based on semiconductor manufacturing processes – promises to bring the benefits of miniaturization to mechanical optical elements: low-cost, high-performance, reliable opto-mechanical components.

MEMS technologies have emerged as a large-scale batch-fabrication approach to making micron- to millimeter-sized mechanical systems. MEMS sensors have been widely available for a variety of different applications, including accelerometers, gyroscopes, and pressure sensors. MEMS actuators are also widely used in inkjet printers, and more recently tracking systems for magnetic disk-drive heads. Research devices are also making their way to commercial markets in a much wider variety of applications including bio-analysis systems, chemical detection systems, and many fiber optic components (tunable lasers, variable-optical attenuators, and fiber-optic switches, among others).

In the same way that semiconductor manufacturing techniques have created a revolution in electronics, so too will the application of these manufacturing techniques create a revolution in opto-mechanical scanners. The miniaturization of these mechanical scanners can result in:
- Reduced size
- Reduced power consumption
- Reduced cost
- Increased speed
- Increased positioning accuracy
- Increased parallelism using arrays
- Increased reliability

MEMS technology can be applied to optical scanning – both resonant beam scanning and steady-state beam steering – and will ultimately result in all of the performance gains mentioned above.

3.   MEMS OPTICAL SCANNERS

Kurt Petersen, in 1980, described the potential use of silicon micromachining to make scanning mirrors [1]. Since then, numerous research groups in both industry and academia have developed scanning mirrors using a variety of fabrication processes. Work from many of these research groups are cited throughout the text.

Work at the University of California, Berkeley has advanced the field of microphotonics under the supervision of Professors Kam Lau, Richard Muller, and Kristofer Pister. This dissertation is focused on demonstrating and mapping the ultimate performance limits of optical-MEMS technology, and is closely related to the work of Jocelyn Nee [2], Matthew Hart [3], Michael Helmbrecht [4], Carl Chang [5], Olav Solgaard, Norman Tien, Mike Daneman [6], and Meng-Hsiung Kiang [7], all of whom worked under the supervision of Professors Richard S. Muller and Kam Lau.

4.   MOTIVATION

MEMS scanners promise to offer improved performance, but with some limitations that make them very attractive for some applications and less attractive for others. These fundamental limitations apply to both conventional and MEMS scanning mirrors, but they have not been widely discussed. The limitations are predicted using well-known mechanical and optical laws. We present them here for engineers from industry and academia to have access to a set of performance targets, and for optical-systems designers to see the ultimate performance limits of MEMS scanning mirrors.

In this research, we have pushed MEMS technology towards high-resolution, high-speed scanning mirrors by using the mechanical and optical performance criteria brought to light in this work. We demonstrate micromirrors made from two fabrication processes to show the advantages and limitations of each particular implementation, thereby giving technologists and researchers additional component-design guidelines and examples of high-speed, high-resolution MEMS scanning mirrors.

5.   OVERVIEW AND ORGANIZATION

The mechanical and optical performance of scanning mirrors can be considered independently of the particular fabrication process chosen, and the ultimate limitations of scanning mirrors are applicable to conventional as well as MEMS mirrors. The mathematical framework for the mechanical and optical performance of scanning mirrors is described in detail in Chapter 2, partially based on previously published results [8].

From the results of Chapter 2 it becomes clear that MEMS-scale mirrors offer significant performance improvements over conventional mirrors for a range of performance parameters – particularly intermediate resolution (100-5000 pixel), high-speed (> 2 kHz) optical scanning and beam steering. Chapter 3 describes the design considerations for a particular implementation of MEMS scanning mirrors: surface-micromachined mirrors, showing that while they may be acceptable for some applications, they cannot satisfy the promise of MEMS scanning mirrors: high-speed, high-performance, low-cost scanners [9].

In Chapter 4 we present an alternative implementation of MEMS scanning-mirror design that does satisfy this promise, partially based on previously published results [10]. This chapter covers the fabrication process, mirror design, and actuator design for this new MEMS scanning mirror, and presents results from a variety of devices made using this process. In this section we also describe results from Jocelyn Nee in a process variation used to make higher-speed, steady-state beam-steering mirrors [11].

In Chapter 5 we show a proof-of-concept demonstration of a raster-scanning video display made using MEMS scanning mirrors [12], [13], [14]. The conclusions of this work are presented in Chapter 6.

Chapter 2

Scanning Mirrors

Scanning mirrors have a wide variety of applications, all of which involve moving the mirror from one position to another, either cyclically or quasi-statically. Resonant scanning – moving the mirror sinusoidally at its resonant frequency – is useful in such applications as line-scanning for barcode readers, raster-scanned video displays, and laser printers. Steady-state beam steering – moving the mirror from one static position to another – is useful for such applications as fiber-optic components (optical switches, variable-optical attenuators, et cetera) and point-to-point optical-communication systems.

This chapter covers the mechanical and optical considerations of scanning mirrors used in both resonant scanning and steady-state beam steering independent of the fabrication technology. These formulations are equally applicable to both macro-scale conventional scanners and MEMS scanners, and will ultimately show the applicability of MEMS-scale mirrors for high-speed (> 5 kHz), moderate-resolution (< 1000 pixel) optical scanning.

There are a number of performance criteria for these two different scanning-mirror modes of operation listed in Table 2-1 together with the various factors that contribute to each, which are listed roughly in order of importance. Where performance criteria are specific to a particular mode of operation (resonant scanning or steady-state beam steering) it is noted in the table.

Table 2-1. Critical performance criteria for scanning mirrors

Performance criteria	Contributing factors
Resonant frequency (only for <u>resonant scanning</u>)	<u>Resonant scanning</u>: Damping, actuator torque
Move time (only for <u>steady-state beam steering</u>)	<u>Steady-state beam steering</u>: Mirror and actuator moment-of-inertia, spring stiffness, actuator torque, damping
Scan angle	<u>Resonant scanning</u>: Actuator torque, damping, hinge strength <u>Steady-state beam steering</u>: Actuator torque, hinge stiffness, hinge strength
Optical resolution	Scan angle, mirror size, mirror static flatness, mirror dynamic flatness
Reflectivity	Optical coating type and thickness
Cost	Package cost, testing cost, fabrication cost, material cost
Power consumption	<u>Resonant scanning</u>: Dynamic actuator efficiency, damping <u>Steady-state beam steering</u>: Static actuator efficiency, dynamic actuator efficiency, damping
Controllability	Integrated sensor resolution, hinge nonlinearities, actuator nonlinearities, damping
Pointing accuracy (only for <u>steady-state beam steering</u>)	<u>Steady-state beam steering</u>: Integrated sensor resolution, thermal sensitivity
Reliability	Hinge material fatigue properties, scan angle, hinge geometry
Robustness (shock survival)	Hinge material properties, hinge geometry, mirror mass

This chapter describes, where possible, the relationships between these performance criteria and the contributing factors. It also provides a qualitative overview of those performance criteria that cannot be mathematically quantified, such as cost.

The mechanical dynamics of scanning mirrors are discussed in section 1, and the optical performance is discussed in section 2. Section 3 summarizes the performance criteria for both resonant scanning and steady-state beam steering.

1. MECHANICAL DYNAMICS

The mechanical dynamics of scanning mirrors is one of the most critical aspects of their design; it determines the optical resolution, scan speed, power consumption, and other performance criteria. Much of the analysis of scanning-mirror dynamics can be performed independently of the details of fabrication process flow or actuator design and can be applied to both macro-scale conventional scanning mirrors and MEMS mirrors. This section presents the foundation for understanding the mechanical dynamics of these mirrors.

1.1 Second-order systems

The dynamics of most scanning mirrors can be modeled as a second-order mechanical system. There are many excellent references on vibrations of mechanical systems [15]; a small subset of the formulations from these sources is presented here to provide background on the analysis of resonant-scanning mirrors.

1.1.1 Linear systems

For a rotational vibratory system, the equation of motion is

$$T = k_\theta \theta + b\dot\theta + I_\theta \ddot\theta \qquad (2\text{-}1)$$

where θ is the mechanical mirror angle, T is the applied torque, k_θ is the torsional stiffness, b is the damping constant, and I_θ is the polar moment-of-inertia about the axis-of-rotation. In general, the torque, stiffness, damping, and moment-of-inertia are all tensors representing different modes of vibration, but in the simple case where the dominant mode is pure rotation about the axis-of-rotation, the torque, stiffness, damping, and moment-of-inertia can be considered scalar values.

Since this idealized system is linear, the response of the system to a nonlinear excitation can be calculated as the superposition of the response to each frequency component of the excitation waveform. The steady-state dynamic response to a sinusoidal driving torque of magnitude T_0 at frequency ω_d is

$$\theta = \theta_0 \sin(\omega_d t + \phi) \qquad (2\text{-}2)$$

where the response scan amplitude θ_0 is given by

$$\theta_0 = \frac{T_0}{\sqrt{\left(k_\theta - I_\theta \omega_d^2\right)^2 + b^2 \omega_d^2}}$$

(2-3)

and the phase ϕ is given by

$$\phi = \tan^{-1}\left(\frac{b\omega_d}{k_\theta - I_\theta \omega_d^2}\right)$$

(2-4)

The amplitude of the response is greatest at the resonant frequency ω_r, which is given by

$$\omega_r = \sqrt{\frac{k_\theta}{I_\theta} - \frac{b^2}{2I_\theta^2}}$$

(2-5)

At this drive frequency, the amplitude is

$$\theta_{0r} = \frac{T_0}{\omega_r b}$$

(2-6)

The quality factor of the resonance Q is simply the amplitude response at the resonant frequency divided by the amplitude response at low frequency. In systems having low damping, the quality factor Q can be approximated as

$$Q \approx \frac{I_\theta \omega_r}{b}$$

(2-7)

and the phase at the resonant frequency is

$$\phi = \tan^{-1}(2Q)$$

(2-8)

1.1.2   Nonlinearities

Many torsional spring designs have significant nonlinearities when driven to large amplitudes, and these nonlinearities in the torsional stiffness make the dynamics of the system somewhat more complex. The spring stiffness can typically be written as

$$k_\theta = k_{\theta 1} + k_{\theta 3}\theta^2 \tag{2-9}$$

For a stiffening spring, $k_{\theta 3}$ is positive, and for a softening spring, $k_{\theta 3}$ is negative. Extensive work has been presented in the literature regarding stiffening-spring dynamics [16] since these nonlinear effects can be quite substantial in MEMS devices.

Nonlinearities are also present if the torque of the actuator is position dependent. For example, if the torque of the actuator is a function of both the mirror angle and an applied drive signal S (which may be proportional to the square of the applied voltage for electrostatic actuators, or to the current for magnetic actuators), and can be represented as

$$T = T(\theta, S) \tag{2-10}$$

then the overall system may be nonlinear with S. The torque can be written as the Taylor's series expansion

$$T = T(\theta_0, S_0) + \frac{dT}{dS}\bigg|_{\theta_0, S_0} S + \frac{d^2T}{dS^2}\bigg|_{\theta_0, S_0} S^2 + ... + \frac{dT}{d\theta}\bigg|_{\theta_0, S_0} \theta + \frac{d^2T}{d\theta^2}\bigg|_{\theta_0, S_0} \theta^2 + ...$$

$$\tag{2-11}$$

If the mirror is held at some bias drive signal S_0 and a small-signal sine wave of amplitude s is applied at frequency ω_d, then the small-signal expansion of the torque (neglecting terms of order two and higher) is

$$T \approx T(\theta_0, S_0) + \frac{dT}{dS}\bigg|_{\theta_0, S_0} s\sin(\omega_d t) + \frac{dT}{d\theta}\bigg|_{\theta_0, S_0} \theta \tag{2-12}$$

The equation of motion becomes

$$T(\theta_0, S_0) + \frac{dT}{dS}\bigg|_{\theta_0, S_0} s \sin(\omega_d t) = \left(k_\theta - \frac{dT}{d\theta}\bigg|_{\theta_0, S_0} \right) \theta + b\dot{\theta} + I\ddot{\theta} \qquad (2\text{-}13)$$

The small-signal resonant frequency at this offset position θ_0 is

$$\omega_{\theta_0} = \sqrt{\frac{k_\theta - \dfrac{dT}{d\theta}\bigg|_{\theta_0, S_0}}{I_\theta} - \frac{b^2}{2I_\theta^2}} \qquad (2\text{-}14)$$

The dependence of the torque on the mirror angle changes the small-signal resonant frequency of the system at different bias points.

1.2   Torsional and lateral spring stiffness

The mechanical design of torsional springs has two major considerations: (1) torsional stiffness, and (2) stiffness in other directions. All of the MEMS mirrors fabricated in this body of work used rectangular cross-section beams to provide at least a portion of restoring force for the mirror.

The torsional stiffness of a pair of rectangular beams is [17]

$$k_\theta = \frac{2G}{L_s} a_h b_h^3 \left[\frac{16}{3} - 3.36 \frac{b_h}{a_h} \left(1 - \frac{b_h^3}{12a_h^4} \right) \right] \qquad (2\text{-}15)$$

where G is the shear modulus of the torsion beam, L_s is the length of the torsion beam, $2a_h$ is the larger of the torsion beam thickness or the width, and $2b_h$ is the smaller of the torsion beam thickness or the width. The strain in the torsion hinge when rotated to an angle θ is

$$\tau_{\max} \approx \frac{3Gb_h}{8L_s} \left[\frac{16}{3} - 3.36 \frac{b_h}{a_h} \left(1 - \frac{b_h^4}{12a_h^4} \right) \right] \theta \qquad (2\text{-}16)$$

The torsion hinge stiffness in the other directions (for example, linear translation along the mirror normal or linear translation perpendicular to the mirror normal and the axis-of-rotation) is an important consideration. Mirror motion in these directions can:

- Add to the effective moment-of-inertia of the mirror, thereby reducing the scan speed;
- Cause unwanted, uncontrolled mirror motions that can affect the position and/or angle of the reflected optical beam;
- Allow large-deflection motions of the mirror that can result in mirror or actuator impact with the substrate and cause particle generation and material wear that can ultimately lead to early-life mirror failure; and
- Cause excessive hinge stresses that may lead to early-life hinge failure.

For rectangular cross-sectioned torsion hinges, any translational motion of the mirror causes both bending and stretching of the torsion hinge. The vertical (along the mirror normal) and lateral (in the in-plane direction perpendicular to the axis-of-rotation and the mirror normal) stiffness of the hinge is nonlinear – the stretching of the torsion hinge with displacement causes a cubic spring term (evident in Equation (2-9)) resulting in greater spring stiffness at large spring displacements. Figure 2-1 shows the orientation of the "vertical" and "lateral" directions relative to the mirror surface and the axis of rotation.

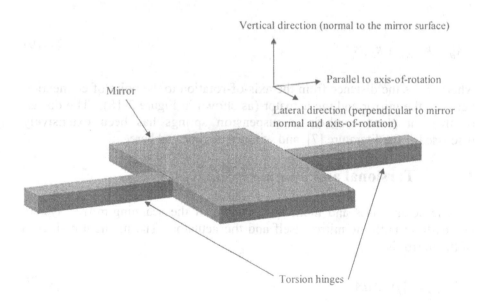

Figure 2-1. Directions of vertical and lateral mirror motion

The linear term of the stiffness, which is accurate only for small-amplitude displacements, underestimates the mirror resonant frequency and overestimates the displacement in these two directions. For design purposes, we can use the linear stiffness to ensure vibration and static displacement in the linear directions is kept within the desired bounds. For this discussion, we will consider only the linear stiffness. The linear stiffness of the pair of torsion hinges in the vertical direction is

$$k_{vert} = \frac{2Et_s^3 w_s}{L_s^3} \qquad (2\text{-}17)$$

The lateral stiffness is

$$k_{lat} = \frac{2Ew_s^3 t_s}{L_s^3} \qquad (2\text{-}18)$$

The surface-micromachined mirrors described in Chapter 3 have an additional spring stiffness exerted by the combdrive suspension. To analyze the dynamics of the surface-micromachined scanning mirrors, the spring stiffness of the suspension k_{lin} can be added to the torsional spring stiffness $k_{torsion}$ as

$$k_\theta = k_{torsion} + k_{lin}H^2 \qquad (2\text{-}19)$$

where H is the distance from the axis-of-rotation to the point of connection between the mirror and the actuator (as shown in Figure 3-18). The design of these in-plane comb-drive suspension springs has been extensively discussed in the literature [7], and will not be repeated here.

1.3   Torsional and lateral resonances

The actual mass and moment-of-inertia of the scanning-mirror system comes from both the mirror itself and the actuator. The moment-of-inertia of the mirror is

$$I_{mirror} = \int \rho tx^2 dA \qquad (2\text{-}20)$$

where x is the distance from the axis-of-rotation to the differential area dA, ρ is the density of the mirror material, and t is the material thickness of the differential area dA. For a rectangular mirror of half length L, half width W, and constant thickness t_m the mirror moment-of-inertia is

$$I_{rect-mirror} = \frac{4\rho\, t_m WL^3}{3} \qquad (2\text{-}21)$$

If the actuator contributes some fraction $\eta_{actuator}$ of this moment-of-inertia, then the total moment-of-inertia is

$$I_{\theta\, rect} = \frac{4\rho t_m L^3 W}{3}\left(1 + \eta_{actuator}\right) \qquad (2\text{-}22)$$

The mass of the rectangular mirror is

$$M_{rect-mirror} = 4\rho\, t_m WL \qquad (2\text{-}23)$$

If the actuator adds some fractional mass $\kappa_{actuator}$, then the total mass is

$$M = 4\rho\, t_m WL\left(1 + \kappa_{actuator}\right) \qquad (2\text{-}24)$$

For scanning mirrors, the resonant frequency of the torsional mode should be significantly lower than the resonant frequencies in the vertical and lateral modes to ensure that there is no significant unwanted displacement in these other modes. Two figures-of-merit for a hinge design are therefore $\dfrac{\omega_{vert}}{\omega_r}$ and $\dfrac{\omega_{lat}}{\omega_r}$. Both of these ratios should be significantly greater than unity to ensure that the mirror motion is torsional rather than translational.

For the rectangular mirror described in this section, the ratio of the vertical and lateral modal frequencies can be calculated for various rectangular cross-section torsion-hinge geometries by combining Equations (2-15), (2-17), (2-18), (2-22), and (2-24):

$$\frac{\omega_{vert}}{\omega_r} = \frac{Lt_s}{L_s b_h}\sqrt{\frac{8(1+\nu)}{3\left[\dfrac{16}{3} - 3.36\dfrac{b_h}{a_h}\left(1 - \dfrac{b_h^4}{12a_h^4}\right)\right]}\left(\frac{1+\eta_{actuator}}{1+\kappa_{actuator}}\right)} \qquad (2\text{-}25)$$

$$\frac{\omega_{lat}}{\omega_r} = \frac{Lw_s}{L_sb_h} \sqrt{\frac{8(1+v)}{3\left[\frac{16}{3} - 3.36\frac{b_h}{a_h}\left(1 - \frac{b_h^4}{12a_h^4}\right)\right]}} \left(\frac{1+\eta_{actuator}}{1+\kappa_{actuator}}\right)1 \qquad (2\text{-}26)$$

If the effects of the actuator (assume $\eta_{actuator} = \kappa_{actuator} = 0$) are ignored,

the ratios $\frac{\omega_{lat}L_s}{\omega_r L}$ and $\frac{\omega_{vert}L_s}{\omega_r L}$ can be plotted using Equations (2-25) and

(2-26) for various ratios of hinge thickness and width, as shown in Figure 2-2. This figure shows that the ratio of the lowest translational-mode resonant frequency (either vertical or lateral) to the torsional resonant frequency is

fixed at $0.977\frac{L}{L_s}$ for all torsional hinge designs (assuming $v = 0.25$),

thereby implying that the hinge should be designed with the constraint

$$L > 1.02\, L_s \qquad (2\text{-}27)$$

Designing with this constraint will ensure that the vertical and lateral resonant mode frequencies are higher than the torsional-mode resonant frequency. This constraint ultimately limits the potential range of mirror designs because the stress in the torsion hinge is inversely proportional to the hinge length. With the constraint of Equation (2-27), the shear stress τ_{max} in the square cross-section hinge due to twisting is

$$\tau_{max} = 0.633\, G^{\frac{3}{4}}\theta\left(\rho\, t_m W\right)^{\frac{1}{4}}\sqrt{\omega} \qquad (2\text{-}28)$$

where G is the shear modulus, θ is the mirror angle of rotation, ρ is the material density, t_m is the thickness of the mirror, W is the mirror width, and ω is the resonant frequency.

For example, with a silicon mirror (G =66 GPa) and maximum shear stress τ_{max} =500 MPa, the maximum angle achievable for a 100 µm-thick by 1 mm-wide mirror with resonant frequency 100 kHz is 3.5 degrees. Interestingly, this limitation is applicable regardless of mirror length. More rigorous analysis of the nonlinear stiffening of the torsion spring will show that for some hinge geometries the nonlinear spring stiffness will prevent large-displacement vibration in the vertical and lateral directions, making the design constraints shown in Equations (2-27) and (2-28) unnecessary.

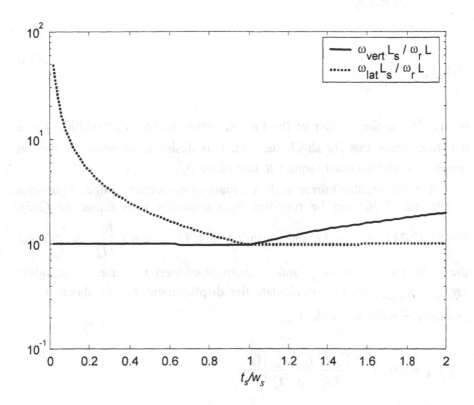

Figure 2-2. Normalized ratios of vertical and lateral resonant frequencies for various ratios of hinge thickness to width (assuming Poisson's ratio 0.25, ignoring mass of actuator, assuming rectangular mirror), calculated from Equations (2-25) and (2-26). Note that in Equations (2-25) and (2-26) $2a_h$ is the larger of t_s and w_s, and $2b_h$ is the smaller of t_s and w_s.

1.4 Mirror mechanical shock

Mechanical shocks to the scanning-mirror system will cause bending of the torsion hinge in one or more of the three linear directions, and this can damage the hinge, resulting in early-life device failure [18]. The stress in the torsion hinge due to this shock is a complex function of the nonlinear spring stiffness, but can be approximated if we assume that the spring stiffness is linear. In this case, the deflection and hinge stress due to acceleration a_{shock} are

$$x_{shock} = \frac{Ma_{shock}L_s^3}{32Ea_hb_h^3} \qquad (2\text{-}29)$$

$$\sigma_{shock} = \frac{3Ma_{shock}L_s}{16a_hb_h^2} \qquad (2\text{-}30)$$

where $2b_h$ is the smaller of the torsion hinge thickness or width. These equations show that for shock survival, it is desirable to make the torsion hinge cross section nearly square to maximize b_h.

For a rectangular mirror with a square-cross-section hinge, Equations (2-29) and (2-30) can be rewritten by combining with Equations (2-5), (2-15), (2-21) and (2-23), and assuming high Q system ($\frac{b^2}{2I_\theta^2} << \frac{k_\theta}{I_\theta}$) and the actuator mass and moment-of-inertia are negligible ($\eta_{actuator}, \kappa_{actuator} << 1$), to calculate the displacement due to shock x_{shock} and the stress due to shock σ_{shock}

$$x_{shock} = 0.623 \frac{(\rho W_m)^{1/4} L_s^{9/4} a_{shock} G^{3/4}}{EL^{5/4} \omega_r^{3/2} t_m^{3/4}} \qquad (2\text{-}31)$$

$$\sigma_{shock} = 2.76 \frac{a_{shock}}{\omega_r} \sqrt{\frac{\rho W_m L_s G}{t_m L^3}} \qquad (2\text{-}32)$$

Equation (2-32) shows that the hinge stress for a given shock is inversely proportional to the resonant frequency. For very low frequency scanning mirrors (like those used in one-dimensional barcode reading, where the resonant frequency is typically about 50 Hz), the hinge stress generated by moderate shock can be large enough to break the torsion hinge. For these low-frequency applications, some over-travel protection may be required to make the scanning mirror robust enough for commercial and industrial use.

1.5   Dynamic deformation

Wave-front aberration is a critical issue for any optical system, and for systems using scanning mirrors, deviations in mirror surface height on the order of the smallest wavelength of interest can significantly reduce the achievable optical resolution (as described more fully in Chapter 2 section 2). Conventional and MEMS technologists alike have expended considerable efforts building mirrors with very flat surfaces [19], [20], [21], [22], [23], and today MEMS and conventional mirrors are available with nonplanarity less than 20 nanometers.

Small static nonplanarity, however, is not sufficient to guarantee diffraction-limited optical performance of resonant scanners. Dynamic deformation of the mirror surface due to inertial loading can lead to significant nonplanarity that can reduce optical performance [24]. In this section, we derive the dynamic deformation of a torsion-bar suspended mirror having an arbitrary width profile as shown in the top view in Figure 2-3a. We assume that the mirror of Figure 2-3a is rotated to some angle θ, and is undergoing an angular acceleration $\ddot{\theta}$. We introduce a Cartesian coordinate system attached to the mirror, as shown in Figure 2-3b such that the x axis is along the undeformed mirror surface perpendicular to the axis of rotation, and y is normal to the undeformed mirror surface. The angular acceleration induces a mirror-surface deformation as shown in Figure 2-3c, where we have rotated the mirror and exaggerated the y axis to show the mirror-surface deformation due to inertial loading. In the following, we derive the mirror displacement due to inertial loading $y(x)$, the deformation (or nonplanarity) of the mirror $d(x)$, and the total peak-to-valley dynamic deformation δ.

Rigorous analysis of this problem requires dynamic analysis of wave propagation in the mirror. The resonant frequency f of the lowest-frequency mode of a rectangular plate oscillating about its center is given by [25]

$$f = 9.82\left(\frac{t_m}{L^2}\right)\left\{\frac{E}{12\rho(1-v^2)}\right\}^{\frac{1}{2}} \tag{2-33}$$

where t_m is the thickness of the mirror, L is its half-length, E is its Young's modulus, v is its Poisson's ratio, and ρ is its density. If the dynamic deformation is small compared to the tilting motion of the mirror, Equation (2-33) predicts a frequency f that is typically orders-of-magnitude higher than the scan frequency. We use this fact to simplify our analysis by considering a static case with the inertial forces as the load.

We assume that the mirror material is isotropic and use equations from plate theory [26] to derive the mirror dynamic deformation. The second derivative of the mirror displacement $y(x)$ with respect to the distance x along the mirror perpendicular to the axis-of-rotation is

$$\frac{d^2 y}{dx^2} = \frac{M(x)}{D(x)} \tag{2-34}$$

where $M(x)$ is the load moment at the point x, and $D(x)$ is the flexural rigidity of the plate, given by

$$D(x) = \frac{E t_m^3 w_m}{12(1 - v^2)} \tag{2-35}$$

where w_m is the mirror width which we take to be x-dependent, and the mirror thickness, density, and material properties are assumed to be constant (see Figure 2-3). The dynamic deformation of the mirror due to inertial loading is symmetrical if the mirror is symmetrical about the axis-of-rotation, so $y(-x) = -y(x)$ and $w(-x) = w(x)$. To simplify the equations presented here, we write the equations for the mirror over the domain $0 \le x \le L$ (where L is the mirror half-length). The load moment at any point $x \ge 0$ is

$$M(x) = \int_x^L F'(x_1)(x_1 - x) dx_1 \tag{2-36}$$

where $F'(x_1)$ is the force-per-unit-length at the point x_1.

There are four forces acting on the mirror: 1) inertial forces, 2) actuator forces, 3) hinge forces, and 4) damping forces. The actuator and hinge forces depend on the specific nature of the attachment between the actuator and the mirror and on the attachment of the mirror to the torsion hinge. For example, the actuator force may be exerted at the extreme edge of the mirror, or the actuator could apply a torque through the torsion hinge itself. The damping forces depend on the ambient fluid viscosity, the mirror shape, and the rate of oscillation of the mirror. For high-Q systems, however, the inertial forces are much higher than the other forces, so we will consider only the effects of the inertial forces on the mirror dynamic deformation.

We assume that the dominant motion is mirror tilt and therefore that the inertial load only depends on the angular acceleration and the position (the additional loading due to the dynamic deformation is negligibly small if the operating frequency is much below the first resonant mode frequency given by Equation (2-33)). The load-per-unit-length at each point on the mirror is the mass-per-unit-length $\rho w_m t_m$ times the acceleration $\ddot{\theta} x$, so

$$F'(x) = -\rho \ddot{\theta} \, w_m t_m x \qquad\qquad (2\text{-}37)$$

where $\ddot{\theta}$ is the mirror angular acceleration, and t is its thickness (see Figure 2-3a).

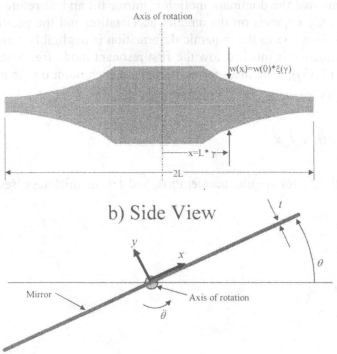

a) Top View

Axis of rotation

$w(x)=w(0)*\xi(\gamma)$

$x=L*\gamma$

2L

b) Side View

t

y

x

θ

Mirror

Axis of rotation

$\dot{\theta}$

c) Rotated and Exaggerated Side View

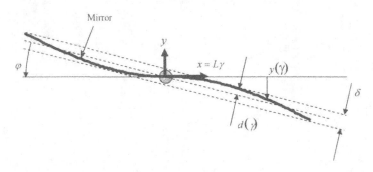

Mirror

y

φ

$x = L\gamma$

$y(\gamma)$

δ

$d(\gamma)$

Figure 2-3. Mirror geometry. a) shows the top view of the mirror, graphically showing variables $x, L, \gamma, w(x), \xi(\gamma)$. b) shows the side of the mirror rotated to the angle θ. c) shows the side view of the mirror rotated by the angle θ and with exaggeration of the y axis to show the mirror dynamic deformation, graphically showing variables $y(\gamma), d(\gamma), \delta$.

Combining Equations (2-34), (2-35), (2-36), and (2-37) gives the mirror displacement as the solution to the equation

$$\frac{d^2 y}{dx^2} = \frac{-12\left(1-v^2\right)\int\limits_x^L \rho\,\ddot\theta\,w_m x_1\left(x_1 - x\right)dx_1}{Et_m^2 w}$$ (2-38)

We introduce the dimensionless variables γ corresponding to the parameterized position, and ξ, corresponding to the parameterized width:

$$\gamma = \frac{x}{L}, -L \le x \le L$$
$$\xi(\gamma) = w(L\gamma), -1 \le \gamma \le 1$$ (2-39)

so that Equation (2-38) becomes

$$\frac{d^2 y}{dx^2} = \frac{-12\rho\,\ddot\theta\,L^3\left(1-v^2\right)}{Et_m^2} \frac{\int\limits_\gamma^1 \xi(\gamma_1)\gamma_1(\gamma_1 - \gamma)d\gamma_1}{\xi(\gamma)}$$ (2-40)

The tilt of the mirror at any point γ is the integral of Equation (2-40)

$$\frac{dy}{dx} = \frac{-12\rho\,\ddot\theta\,L^4\left(1-v^2\right)}{Et_m^2} \int\limits_0^\gamma \frac{\int\limits_{\gamma_2}^1 \xi(\gamma_1)\gamma_1(\gamma_1 - \gamma_2)d\gamma_1}{\xi(\gamma_2)}d\gamma_2$$ (2-41)

Integrating Equation (2-41) gives the mirror displacement $y(\gamma)$ resulting from inertial forces

$$y(\gamma)=\frac{-12\rho\ddot{\theta}\,L^5\left(1-v^2\right)}{Et_m^2}\int_0^\gamma\left[\int_0^{\gamma_3}\frac{\int_{\gamma_2}^1\xi(\gamma_1)\gamma_1(\gamma_1-\gamma_2)d\gamma_1}{\xi(\gamma_2)}d\gamma_2\right]d\gamma_3,\quad 0\leq\gamma\leq1$$

(2-42)

For a rectangular mirror $\xi(\gamma)=1$, so the mirror displacement for a rectangular mirror $y_{rect}(\gamma)$ from Equation (2-42) is

$$y_{rect}(\gamma)=\frac{-12\rho\ddot{\theta}\,L^5\left(1-v^2\right)}{Et_m^2}\left(\frac{1}{6}\gamma^2-\frac{1}{12}\gamma^3+\frac{1}{120}\gamma^5\right)$$

(2-43)

Figure 2-4a is a plot of the displacement of a rectangular mirror according to Equation (2-43) (the displacement for $-1\leq\gamma<0$ is plotted as $-y_{rect}(|\gamma|)$). Figure 2-4a shows that the inertial loading causes a tilt φ in the direction opposite to the angular acceleration. This tilt from inertial loading can be large enough to cause a shift of the beam reflected from the mirror, and should be taken into account for high-resolution scanning-mirror design.

In this derivation, the mirror tilt from inertial loading φ is important in calculating the peak-to-valley dynamic deformation δ. The total displacement from the undeformed mirror surface from Equation (2-43) is $2.2\dfrac{\rho\ddot{\theta}\,L^5\left(1-v^2\right)}{Et^2}$ (between mirror ends). However, the important quantity to characterize optical distortion as a result of dynamic deformation is the deviation from planarity rather than the total displacement across the mirror. To calculate the deviation from planarity, we must rotate the coordinate system by the mirror tilt φ. Since we consider only small dynamic deformation, we can approximate this coordinate-system rotation by subtracting $\Delta y=\gamma L\tan(\varphi)$ from $y(\gamma)$ in Equation (2-42) to get an expression for the mirror deformation $d(\gamma)$ as shown in Figure 2-3c.

$$d(\gamma) = \frac{-12\rho\,\ddot{\theta}\,L^5\left(1-v^2\right)}{Et_m^2}\left\{\int_0^\gamma\left[\int_0^{\gamma_3}\frac{\int_{\gamma_2}^1 \xi(\gamma_1)\gamma_1(\gamma_1-\gamma_2)d\gamma_1}{\xi(\gamma_2)}d\gamma_2\right]d\gamma_3\right\}$$

$$-L\gamma\tan(\varphi)$$

(2-44)

We can rewrite Equation (2-44) by moving the $\tan(\varphi)$ term inside the brackets and replacing $L\tan(\varphi)$ with the slope Ψ

$$d(\gamma) = \frac{-12\rho\,\ddot{\theta}\,L^5\left(1-v^2\right)}{Et_m^2}\left\{\int_0^\gamma\left[\int_0^{\gamma_3}\frac{\int_{\gamma_2}^1 \xi(\gamma_1)\gamma_1(\gamma_1-\gamma_2)d\gamma_1}{\xi(\gamma_2)}d\gamma_2\right]d\gamma_3-\Psi\gamma\right\}$$

(2-45)

where the slope Ψ is related to the mirror tilt due to inertial loading φ by

$$\Psi = \frac{Et_m^2}{12\rho\,\ddot{\theta}L^4\left(1-v^2\right)}\tan(\varphi)$$ (2-46)

To calculate the peak-to-valley dynamic deformation δ, we choose the slope Ψ so that the mirror surface deviates from the planarity by equal amounts in both the positive and negative directions; stated mathematically

$$\max[d(\gamma),0\le\gamma\le1]=-\min[d(\gamma),0\le\gamma\le1]$$ (2-47)

Geometrically, this mathematical expression is equivalent to drawing the three parallel lines shown in Figure 2-3c. The slope of these lines is Ψ, the center line goes through the axis-of-rotation, and the other two lines are spaced as far as possible while still intersecting the mirror surface. The total peak-to-valley surface deformation δ is the distance between the two parallel lines bordering the mirror excursions from planarity, which can be calculated as

$$\delta = \left|\max[d(\gamma), 0 \leq \gamma \leq 1] - \min[d(\gamma), 0 \leq \gamma \leq 1]\right| = 2\left|\max[d(\gamma), 0 \leq \gamma \leq 1]\right|$$

(2-48)

For a rectangular mirror $\xi(\gamma) = 1$ and Equation (2-45) reduces to

$$d_{rect}(\gamma) = \frac{-12\rho\ddot{\theta} L^5 (1 - v^2)}{E t_m^2} \left(\frac{1}{6}\gamma^2 - \frac{1}{12}\gamma^3 + \frac{1}{120}\gamma^5 - \Psi\gamma\right)$$

(2-49)

We can numerically calculate the maximum and minimum of Equation (2-49) as a function of Ψ, and then choose Ψ to satisfy Equation (2-47). For the rectangular mirror, this results in

$$d_{rect}(\gamma) = \frac{-12\rho\ddot{\theta} L^5 (1 - v^2)}{E t_m^2} \left(\frac{1}{6}\gamma^2 - \frac{1}{12}\gamma^3 + \frac{1}{120}\gamma^5 - 0.0803\gamma\right)$$

(2-50)

Figure 2-4b shows the surface deformation from Equation (2-50). The peak-to-valley dynamic deformation is calculated using Equation (2-48). The mirror deformation $d_{rect}(\gamma)$ has a maximum at $\gamma = 1$, so plugging into Equation (2-50) and multiplying by 2 (as described in Equation (2-48)) gives the peak-to-valley dynamic deformation δ_{rect}

$$\delta_{rect} = \frac{0.272\rho\ddot{\theta} L^5 (1 - v^2)}{E t_m^2}$$

(2-51)

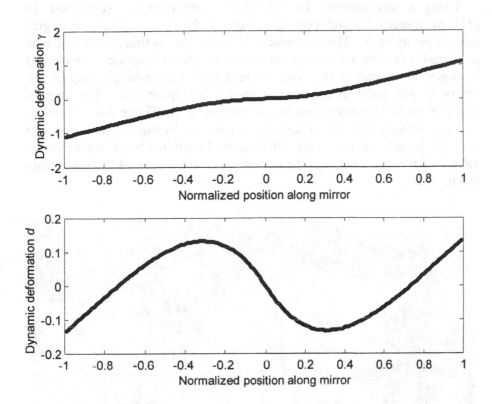

Figure 2-4. Calculated dynamic deformation of rectangular mirror of constant cross section. Top figure shows deformation $y(\gamma)$ from Equation (2-43). Bottom figure shows deformation $d(\gamma)$ from Equation (2-50). (Note: the dynamic deformation shown in the figure has been scaled by $\dfrac{\rho\ddot{\theta}L^{5}\left(1-v^{2}\right)}{Et_{m}^{2}}$; displacement $y(\gamma)$ and deformation $d(\gamma)$ for $-1\leq\gamma<0$ are plotted as $-y(|\gamma|)$ and $-d(|\gamma|)$ respectively).

1.6 Experimental verification

To verify the dynamic-deformation calculations derived above, we have fabricated a MEMS scanning mirror using the fabrication process described in Chapter 4 section 1. Figure 2-4 shows a scanning-electron micrograph (SEM) of the mirror used for this study. This micromachined mirror is made of single-crystal silicon, and is actuated with integrated torsional electrostatic actuators. The mirror surface is very nearly planar in its rest state – its nonplanarity is lower than 10 nm across the entire surface of the mirror.

Using a stroboscopic Twyman-Green interferometer (described in [27]),we measured a surface-height map for the mirror at various angles throughout its scan. The mirror scan is sinusoidal in time, so the angular acceleration (and thus the nonplanarity resulting from dynamic deformation) is largest at the end of the scan. A measured surface-height map of the mirror at this point in the scan is shown in Figure 2-6. The angular acceleration of the mirror when the measurements for Figure 2-6 were taken was 2.31 radians per second squared. Using the Young's modulus of 130 MPa [28], and Poisson's ratio of 0.22 in Equation (2-50), we calculate values for the deformation of the mirror which are compared to experiments in Figure 2-7.

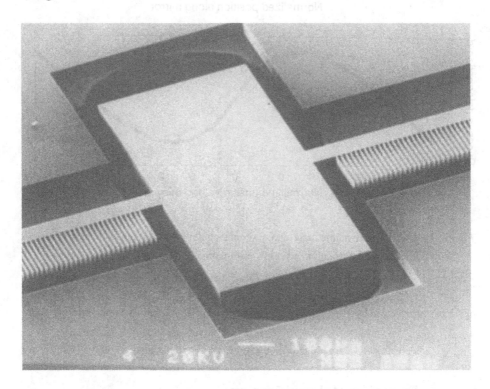

Figure 2-5. Scanning-electron micrograph of MEMS scanning mirror

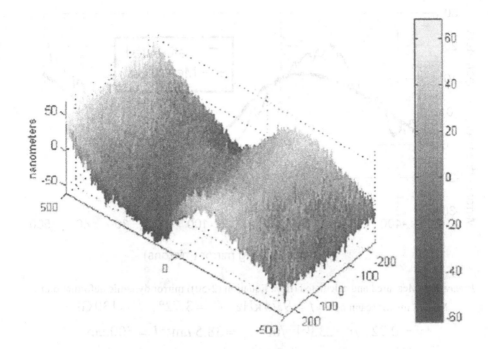

Figure 2 6. Interferogram of dynamic mirror surface deformation when the mirror is at the
maximal scan angle.

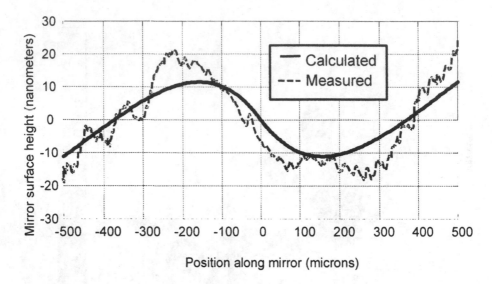

Figure 2-7. Measured and calculated (from Equation (2-50)) mirror dynamic deformation for MEMS mirror scanning at $f = 9.24\,\text{kHz}$, $\theta_0 = 3.92°$, $E = 130\,\text{GPa}$,

$$v = 0.22, \; \rho = 2330\,\text{kg/m}^3, \; t_m = 38.5\,\mu\text{m}, \; L = 500\,\mu\text{m}.$$

1.7   Non-rectangular mirror shapes

The dynamic deformation can be reduced by using non-rectangular mirror-width profiles like that shown schematically in Figure 2-3a. . For example, elliptical and diamond-shaped mirrors have lower mass in locations towards the mirror tip, and therefore exhibit lower dynamic deformation than rectangular mirrors.   In this section, we consider the dynamic deformation of mirrors having arbitrary width-profiles. To do this, we introduce a dynamic deformation factor r that depends only on the mirror-width profile. We define r such that the dynamic deformation of a mirror having an arbitrary width profile is written in comparison to a rectangular mirror so that from Equation (2-51)

$$\delta = \frac{0.272\rho\,\ddot{\theta}\,L^5\left(1 - v^2\right)}{E t_m^2} r \qquad\qquad (2\text{-}52)$$

To calculate r for an arbitrary mirror-width profile, we define the term in the brackets of Equation (2-45) as $c(\gamma)$,

$$c(\gamma) = \int\limits_{0}^{\gamma}\left[\int\limits_{0}^{\gamma_3}\frac{\gamma_2\int\limits_{0}^{1}\xi(\gamma_1)\gamma_1(\gamma_1-\gamma_2)d\gamma_1}{\xi(\gamma_2)}d\gamma_2\right]d\gamma_3 - \Psi\gamma \qquad (2\text{-}53)$$

As we did for the rectangular mirror, we choose Ψ to remove the tilt resulting from inertial loading to calculate the mirror deviation from planarity. The tilt Ψ satisfies the constraint

$$\max[c(\gamma), 0 \leq \gamma \leq 1] = -\min[c(\gamma), 0 \leq \gamma \leq 1] \qquad (2\text{-}54)$$

The dynamic-deformation factor r is calculated for arbitrary mirror-width profiles using Equations (2-48), (2-53), and (2-54). The resulting dynamic-deformation factor r is

$$r = \frac{24}{0.272}\left|\max[c(\gamma), 0 \leq \gamma \leq 1]\right| \qquad (2\text{-}55)$$

For a rectangular mirror, the solution to Equations (2-53), (2-54), and (2-55) yields $r = 1$. The resulting deformation δ of an arbitrary mirror-width profile $\xi(\gamma)$ is given by Equation (2-52).

Using Equations (2-53), (2-54), and (2-55) we can calculate the deformation factor r for an arbitrary mirror-width profile, and we can numerically evaluate different mirror-width profiles to minimize r in order to find the optimal width profile. Only solutions for practical mirror-width profiles need to be considered.

One example of an impractical mirror design with little dynamic deformation is that of a mirror that is very narrow at the axis-of-rotation. In the extreme case with zero width at the axis-of-rotation, applying a torque will not cause the rest of the mirror to rotate at all, so there will be no dynamic deformation and no optical scan. In order to eliminate this uninteresting solution to the optimal mirror-width profile, we place an additional constraint that the mirror width must decrease monotonically from the axis of rotation outward.

In addition, with most fabrication processes there is some limitation to the minimum mirror width at any point. To calculate how this minimum width affects the mirror dynamic deformation, we constrain the mirror width such that its tip width is some fraction κ times its width at the axis-of-rotation; i.e. the normalized width ξ is limited by

$$\xi(\gamma) \geq \kappa \, \xi(0) \quad -1 \leq \gamma \leq 1 \tag{2-56}$$

Using numerical techniques, we optimize $\xi(\gamma)$ to minimize the deformation factor r and thus calculate the optimal mirror-width profile. To perform this optimization, we approximate the continuous mirror-width profile as a set of discrete constant-width sections. The integrals in Equation (2-53) are performed numerically to calculate $c(\gamma)$. Then Ψ is chosen to satisfy Equation (2-54), and the deformation factor r is calculated using Equation (2-55). The mirror widths at these discrete sections are then adjusted using standard minimization routines to reduce the deformation factor r, and these procedures are repeated until further changes in the width profile do not significantly affect the deformation factor r. This optimization procedure is repeated for various values of the tip width constraint κ.

In Figure 2-8 we show the calculated optimal width profile as obtained by these procedures for $\kappa = 0$, and in Figure 2-9 we show the calculated optimal width profile for $\kappa = 0.1$. Figure 2-9 shows that for $\kappa = 0.1$ the mirror width at the center (where the torsion hinge is attached) is widest and it narrows to 10% of the maximum width at its tip.

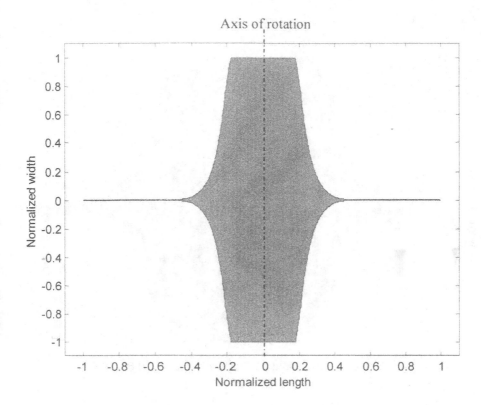

Figure 2-8. Calculated optimal mirror width profile for $\kappa = 10^{-8}$. Torsional hinges are assumed to be attached at the top and bottom of the mirror along the axis-of-rotation.

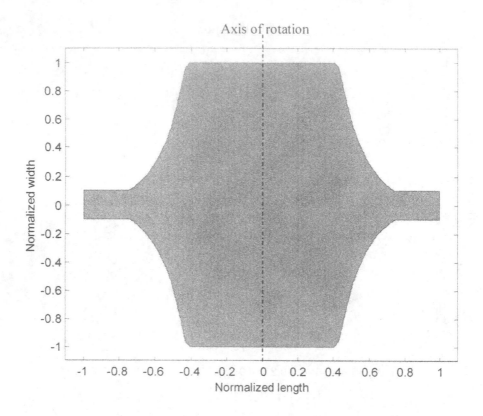

Figure 2-9. Calculated optimal mirror width profile for $\kappa = 0.1$. Torsional hinges are assumed to be attached at the top and bottom of the mirror along the axis-of-rotation.

Table 2-2 shows the deformation factor r for various mirror-width profiles, indicating that dynamic deformation can be reduced to less than 1% of the rectangular mirror dynamic deformation with an extremely narrow mirror tip; r can be reduced by a lesser, still significant amount with more practical limitations. Figure 2-10 shows the deformation factor r for various values of κ.

Table 2-2. Deformation factor r for various mirror-width profiles.

Shape	Deformation factor r
Rectangle	1
Ellipse	0.560
Diamond	0.234
Calculated optimal shape, $\kappa = 0.1$	0.203
Calculated optimal shape, $\kappa = 0.01$	0.0796
Calculated optimal shape, $\kappa = 10^{-8}$	0.0075

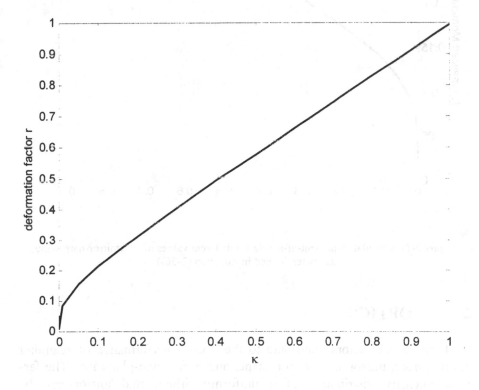

Figure 2-10. Calculated deformation factor r for different values of κ (mirror-narrowing parameter defined in Equation (2-56)), using Equations (2-53), (2-54), and (2-55).

Reducing the deformation factor r has the additional benefit of reducing the mirror moment-of-inertia, as shown in Figure 2-11. Reduced moment-of-inertia allows faster static mirror positioning, and has the added benefits that it reduces stress in the torsional hinges. Reduced moment-of-inertia also reduces the energy stored in the system thereby making smaller the resonant quality factor Q and allowing high-amplitude mirror motions over a larger range of frequencies.

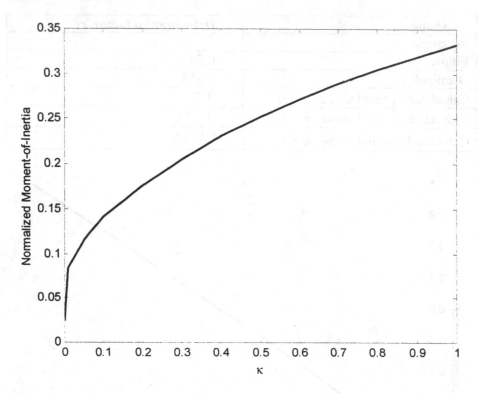

Figure 2-11. Calculated moment-of-inertia for different values of κ (mirror-narrowing parameter defined in Equation (2-56)).

2.   OPTICS

Three major factors contribute to the optical performance of scanning micromirrors: mirror size, mirror shape, and mirror non-planarity. The far-field intensity distribution of a uniformly illuminated mirror can be calculated using the Fourier transform, as described in [29]. The aperture function $A(\gamma)$, is calculated from the mirror deformation $y(\gamma)$ over the mirror aperture

$$A(\gamma) = \begin{cases} j2\pi\dfrac{2y(\gamma)}{\lambda}, & 0 \le \gamma \le 1 \\[2mm] -j2\pi\dfrac{2y(-\gamma)}{\lambda}, & -1 \le \gamma \le 0 \\[2mm] 0, & |\gamma| > 1 \end{cases} \tag{2-57}$$

where $j = \sqrt{-1}$. The far-field intensity distribution is proportional to the square of the Fourier transform of this aperture function.

Fourier optics (valid when the distance to the viewing screen is much larger than the size of the mirror in question) can be used to analyze the effects of nonplanar mirror surfaces and diffraction from the mirror aperture to determine the optical performance of a scanning mirror.

2.1   Mirror size and beam divergence

Figure 2-12 shows the calculated far-field intensity distribution for a three planar mirrors of different sizes. The reflected beam width shows significantly more optical divergence for small mirrors than for large mirrors. The beam divergence can be quantified using a number of different criteria; the choice of the appropriate criteria is often dependent on the specific application. For video displays, it is customary to use the full-width-half maximum beam divergence, which corresponds to a Modulation Transfer Function (MTF) of 50%. The full-width-half-max beam divergence from a uniformly-illuminated aperture of size $2L$ is

$$\beta = \frac{a\lambda}{2L} \tag{2-58}$$

where the aperture shape factor a depends on the shape of the aperture. The aperture shape factor a is calculated using the full-width-half-max angular divergence of the far-field intensity distribution calculated using the Fourier-transform technique described above and in [29]. For a rectangular mirror, $a = 1$.

The beam divergence orthogonal to the scan can be calculated by replacing the mirror half-length with the mirror half-width. For a mirror scanning $\pm\theta_0$ mechanical, the total optical scan angle is somewhere between $2\theta_0$ and $4\theta_0$, depending on the angle between the incoming beam and the axis of mirror rotation [30]. The largest scan occurs when the incoming beam is perpendicular to the axis-of-rotation, in which case the total optical scan angle is $4\theta_0$. The resolution of the system N_{pixels} in this case is

$$N_{pixels} = \frac{8\theta_0 L}{a\lambda} \tag{2-59}$$

The resolution here is reported in "pixels", but this is not completely analogous to the "pixels" of liquid-crystal displays. Since the optical scan is contiguous, the optical spot generated at each instant is convolved with its neighboring spots to form a contiguous optical line. The modulation rate of the light source is not limited to this resolution, but the number of resolvable optical spots is equivalent to the number of "pixels" in Equation (2-57).

Various optical systems placed after the scanning mirror can be used to increase or decrease the optical scan angle, but the product of the scan angle θ and the mirror aperture $2L$ is a Lagrange invariant (assuming the same index-of-refraction on both sides), i.e.

$$\theta L = \theta' L' \tag{2-60}$$

The total optical resolution, therefore, is fixed independently of any optical system placed after the scanning mirror. This fact is useful because it decouples the mechanical scan angle of the mirror from the desired angular optical range while maintaining the same resolution.

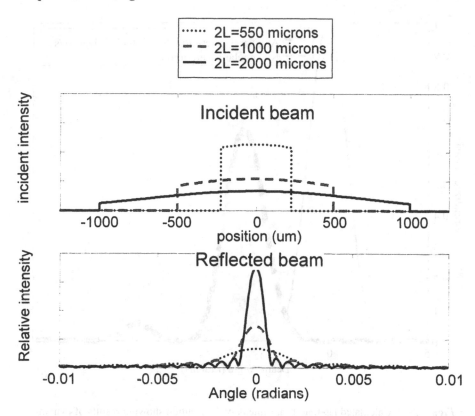

Figure 2-12. Beam divergence for three rectangular mirror lengths (incident light is 635 nm).

2.2 Dynamic deformation effects on optical resolution

The beam divergence from a mirror subject to angular acceleration is due to both diffraction and mirror dynamic deformation. Figure 2-13 shows the calculated far-field intensity distribution of a uniform-intensity optical beam reflected from a rectangular mirror with no angular acceleration and from the same mirror with angular acceleration using the surface height y as calculated in Equation (2-43). (Since the surface height described by Equation (2-43) has finite tilt φ, the beam center is shifted, in this case by approximately 4.8×10^{-4} radians; if Equation (2-50) were used, the beam would be centered in the figure.) The angular acceleration clearly causes some beam spreading, which is evident in the reduced intensity in the main intensity peak, and the non-negligible optical intensity in the additional peaks.

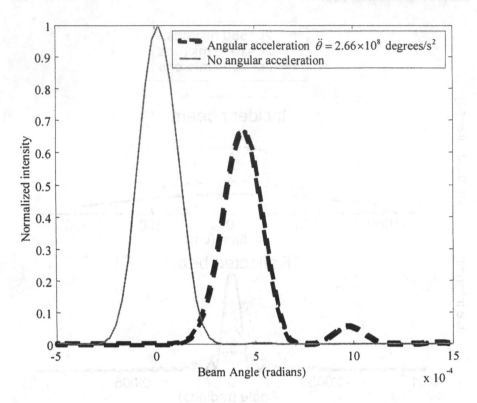

Figure 2-13. Calculated far-field beam-intensity distribution showing results of dynamic deformation. ($L = 455\,\mu m$, $t_m = 100\,\mu m$, $\rho = 2330\,kg/m^3$, $E = 168.9\,GPa$, $\nu = 0.22$, $\lambda = 635\,nm$).

Consideration of the Rayleigh limit provides a useful starting point for an analysis of the optical effects of mirror dynamic deformation. Rayleigh's limit states that if the wave-front aberration is smaller than $\lambda/4$, the optical resolution is nearly diffraction limited. Since the wave-front aberration of a plane wave reflected from a mirror is twice the deviation from planarity over the mirror surface, the Rayleigh limit constrains the total (static plus dynamic) peak-to-valley mirror deviation from planarity to lower than $\lambda/8$ to retain nearly diffraction-limited optical performance.

Static deformations can often be compensated by using lenses, but dynamic deformation changes with time and is therefore impossible to correct with static optical elements. If we assume that static deformation is either negligible or compensated, we can use our theory to calculate the largest mirror length capable of diffraction-limited optical performance as a function of the maximum angular acceleration by using Equation (2-52).

Setting the peak-to-valley dynamic deformation equal to the Rayleigh limit $\delta = \dfrac{\lambda}{8}$ and solving for the angular acceleration $\ddot{\theta}$ gives

$$\ddot{\theta} = \frac{Et^2}{0.272\rho L^5\left(1-v^2\right)r}\frac{\lambda}{8} \tag{2-61}$$

If the mirror scan is sinusoidal with radial frequency ω, then the angular acceleration is

$$\ddot{\theta} = \omega^2\theta_0 \tag{2-62}$$

where θ_0 is the amplitude of the sinusoidal motion (the half-angle mechanical scan angle). Combining Equations (2-61) and (2-62) gives the maximum resonant frequency $f = \dfrac{\omega}{2\pi}$ of the diffraction-limited scanning mirror

$$f = \frac{1}{2\pi}\sqrt{\frac{Et_m^2}{0.272\rho L^5\left(1-v^2\right)r\theta_0}\frac{\lambda}{8}} \tag{2-63}$$

Equations (2-59) and (2-63) can be combined to give the maximum resonant frequency for a particular desired optical resolution

$$f = 19.53\frac{t_m\theta_0^2}{\lambda^2}\sqrt{\frac{E}{\rho\left(1-v^2\right)N_{pix}^5\, a^5 r}} \tag{2-64}$$

Equation (2-64) shows that the frequency for a particular desired resolution can be increased by increasing the mirror scan angle θ_0 or the mirror thickness t_m. Figure 2-14 shows a plot of the highest possible resolution for rectangular mirrors of three thicknesses scanning at various frequencies, based on Equation (2-64), and Figure 2-15 shows the corresponding mirror lengths as calculated from Equation (2-59). In both cases, the scan angle θ_0 is assumed to be 10°.

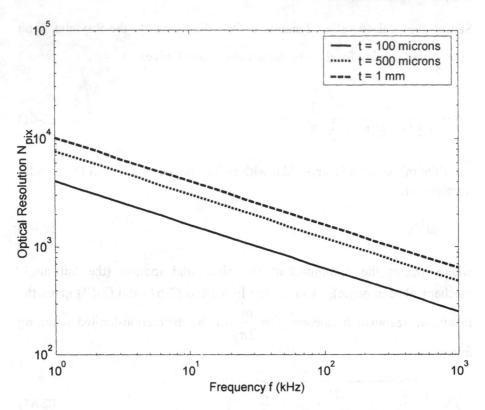

Figure 2-14. Highest frequency for desired optical resolution and mirror thickness calculated from Equation (2-64), assuming rectangular mirror ($r = 1$, $a = 1$), single-crystal silicon mirror ($\rho = 2330\,\text{kg/m}^3$, $E = 168.9\,\text{GPa}$, $\nu = 0.22$), scan angle of $\theta_0 = 10°$. wavelength $\lambda = 635\,\text{nm}$.

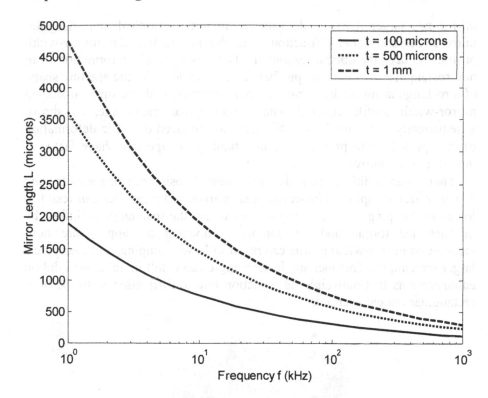

Figure 2-15. Optimal mirror half length for various resonant frequencies calculated using Equations (2-64) and (2-59), assuming rectangular mirror ($r = 1, a = 1$), single-crystal silicon mirror ($\rho = 2330 \, \text{kg/m}^3$, $E = 168.9 \, \text{GPa}$, $v = 0.22$), scan angle of $\theta_0 = 10°$, wavelength $\lambda = 635 \, \text{nm}$.

The dynamic deformation factor r and the aperture shape factor a for various mirror geometries are shown in Table 2-3. The non-rectangular mirror shapes listed in Table 2-3 show reduced dynamic deformation, but the aperture shape factor a is larger for non-rectangular mirrors. The ultimate purpose of the different mirror shapes is to allow higher-frequency operation for a particular mirror thickness and desired optical resolution; however the increased aperture shape factor a for non-rectangular mirrors serves to reduce the optical resolution, as shown in Equation (2-59). This reduces the achievable resonant frequency for a particular optical resolution N_{pixels}, as shown in Equation (2-64).

Equation (2-64) shows that the maximum frequency is proportional to the geometry factor $\sqrt{\dfrac{1}{a^5 r}}$. Table 2-3 shows that although the rectangular mirror has larger dynamic deformation than the elliptical and triangular mirrors, the additional diffraction from the non-rectangular mirror-width profiles reduces the optical resolution. The low dynamic deformation from non-rectangular mirror-width profiles can be coupled with the aperture shape of a rectangular mirror by using a support structure with the optimal-shaped mirror-width profile covered with a rectangular membrane, as shown schematically in Figure 2-16. This gives the reduced dynamic deformation of the optimal width profile while maintaining the aperture shape factor of the rectangular mirror.

There is an additional consideration when choosing mirror geometry and that is mirror damping. Non-rectangular mirror-width profiles can result in lower air damping. For scanning mirrors where the scan angle is limited by the actuator torque and air damping, choosing an appropriate non-rectangular mirror-width profile can result in lower damping, and therefore a larger scan angle. This increase in scan angle can result in optical-resolution enhancements that outweigh the reduction due to diffraction from the non-rectangular optical aperture.

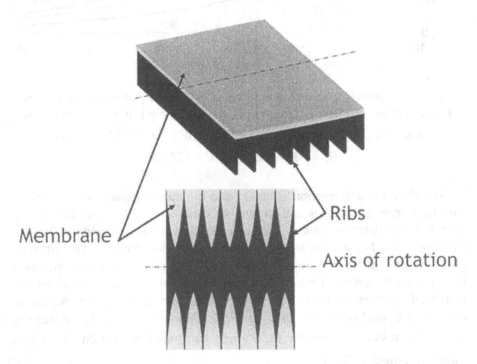

Figure 2-16. Schematic of optimum-width-profile mirror with rectangular membrane optical surface.

Table 2-3. Maximum resonant scan frequency for various mirror shapes.

Shape	Deformation factor r	Aperture shape factor a	Normalized maximum $\dfrac{f}{f_{rectangle}}$ frequency
Rectangle	1	1	1
Ellipse	0.560	1.25	0.76
Diamond	0.234	1.7	0.54
Calculated optimal profile for dynamic deformation, κ =0.1, with rectangular membrane optical surface	0.203	1	2.22

From the above considerations, we can describe the relationships between the contributing factors and the performance criteria shown in Table 2-1. This section refers to the equations derived earlier to summarize the important performance criteria for both resonant-scanning and steady-state beam-steering mirrors. First we will consider those criteria common to both resonant-scanning and steady-state beam-steering mirrors, and then we will consider criteria distinct to each of the two types of scanning mirrors.

3. SUMMARY OF PERFORMANCE CRITERIA

From the above considerations, we can describe the relationships between the contributing factors and the performance criteria shown in Table 2-1. This section refers to the equations derived earlier to summarize the important performance criteria for both resonant-scanning and steady-state beam-steering mirrors. First we will consider those criteria common to both resonant-scanning and steady-state beam-steering mirrors, and then we will consider criteria distinct to each of the two types of scanning mirrors.

3.1   Common criteria

Optical Resolution
The optical resolution is limited by the mirror size, mirror shape, and scan angle, as shown in Equation (2-59).

Reflectivity
The reflectivity depends on the fabrication process chosen and is ultimately limited by cost and technology concerns rather than by fundamental factors.

Cost
The cost is dictated by technology and industrial limitations rather than by technical limitations. For MEMS mirrors, the cost often correlates with the overall chip size, which can be estimated based on the mirror length, mirror width, and hinge length, but the exact size often depends on the actuator geometry, the fabrication process flow, and the packaging.

Reliability
The reliability is ultimately determined by the hinge material fatigue properties and hinge stress, which can be calculated from Equation (2-16). Package reliability can also be a critical factor in scanning mirror reliability.

Robustness
The robustness of the mirror is determined by the hinge material strength and the stress due to any applied shock, given by Equation (2-30). Another contributor to robustness is the allowable mirror deflection before contact between the moving mirror and actuator or substrate occurs. This limitation can be quite severe for electrostatic combdrives often used to actuate MEMS mirrors because of the small electrostatic gap that is desirable to increase the actuator force. Equation (2-29) can be used as a conservative estimate of the deflection due to an external shock.

3.2   Resonant scanners

Resonant frequency
The ultimate resonant frequency for a resonant scanning mirror with diffraction-limited optical performance is shown in Equation (2-64). The resonant frequency can be chosen based on the application, and the hinge geometry can be adjusted to ensure the desired resonant frequency.

Scan angle

The scan angle achievable with an actuator torque T_0 and desired resonant frequency ω_r is given by Equation (2-6). If the scan angle is limited by other optical considerations (like the numerical aperture of the optical system), Equation (2-6) can be used to determine the necessary torque.

Power consumption

The power consumption of the resonant-scanning mirror is

$$P = \frac{1}{2} b \omega_r^2 \theta_0^2 \tag{2-65}$$

Controllability

High-Q resonant-scanning mirrors have very reliable sinusoidal scans which are impervious to distortion due to the high energy storage. Distortions in the scan can be reduced by increasing the quality factor by either increasing the moment-of-inertia or decreasing the damping. The scan angle and resonant frequency can be maintained by increasing the mirror thickness while maintaining the mirror lateral dimensions.

3.3 Steady-state beam-steering mirrors

Move time

The move time for a steady-state beam-steering mirror is related to the quality of the control algorithms used to reposition the mirror, the torque applied to the mirror, the spring stiffness, and the damping. In general, the move time from position θ_1 to θ_2 depends on both the excess accelerating torque ($T_+ = T_{max} - \dfrac{\theta_2}{k_\theta}$), and the decelerating torque ($T_- = T_{min} - \dfrac{\theta_2}{k_\theta}$). If we ignore the change in spring torque over the course of the scan (which is a good assumption for small moves), and ignore the effects of damping, then the minimum move time is

$$t_{move} \approx \sqrt{\frac{2I_\theta(\theta_2 - \theta_1)\left(1 - \dfrac{T_+}{T_-}\right)}{T_+}} \tag{2-66}$$

where T_- is a negative number. This shows that the move time is determined by the maximum and minimum actuator torque and the mirror moment-of-inertia I_θ. If the system is low Q the damping can increase the move time, and if the system is high Q, oscillations in the mirror response may increase the mirror settling time beyond that shown in Equation (2-57).

Scan angle

The scan angle achievable with an actuator torque T_0 is given by Equation (2-3). If the scan angle is limited by other optical considerations (like the numerical aperture of the optical system), then Equation (2-3) can be used to determine the necessary torque.

Power consumption

The power consumption of steady-state beam-steering mirrors depends on both the dynamic and static power consumption. For some actuator designs, the steady-state power consumption can be quite significant (for example, magnetic actuators). The lower bound power consumption for steady-state beam steering is zero. However, any practical actuator will have some finite power consumption in the rest position, although it can be quite small for some actuators (such as electrostatic).

Controllability

For steady-state beam steering mirrors, it is desirable to lower the moment-of-inertia to improve controllability. This reduces the quality factor of the resonance Q and reduces the effects of wideband noise on the system. Integrated high-accuracy, low-noise position sensors can also be helpful in building an easily-controllable steady-state beam steering mirror. High-quality commercially available galvanometric scanners have closed-loop position feedback using capacitive or optical sensors, and some MEMS scanning mirrors have integrated sensors that can be used for position feedback.

Chapter 3

Surface-Micromachined Mirrors

Surface micromachining is a set of fabrication processes derived from planar integrated-circuit fabrication technologies that have been adapted to make free-standing mechanical structures. Surface micromachined devices were made in the late 1960's by Nathanson [31], and the foundations for the polysilicon surface-micromachining process were laid by Muller and Howe in 1982 [32]. Steps in the fabrication process include, but are not limited to:

- Polycrystalline silicon (polysilicon) thin-film low-pressure chemical-vapor deposition (LPCVD). Doped and undoped films can be deposited in thicknesses that range between hundreds of nanometers to tens of microns.
- Silicon nitride LPCVD.
- Silicon dioxide LPCVD.
- Thermal oxidation. Both wet and dry thermal oxidation of single-crystal and polysilicon are used to create silicon dioxide layers.
- Photolithography. Spin-on photoresist is used with contact-mask or stepper photolithography to produce patterns in the photoresist that can be used to delineate etch masks for thin-film etching.
- Wet etching. Hydrofluoric acid (HF) etching is used to etch silicon dioxide without significantly eroding silicon nitride, single-crystal silicon, or polysilicon.
- Dry etching. Plasma etching is used to selectively remove silicon, silicon nitride, and silicon dioxide.
- Metal evaporation. Both evaporation and sputtering of gold or aluminum are used to form pads for electrical contact and for enhancing the reflectivity of mirror surfaces.

These steps are typically combined in surface micromachining to produce planar silicon structures with interspersed silicon dioxide sacrificial layers that are etched away during the final release steps to produce free-moving structures. Table 3-1 shows a simplified version of the process flow for the two-structural layer polysilicon surface-micromaching process provided by the Cronos Multi-User Micromachining Process (MUMPS). A schematic cross-section of a device made in this process is shown in Figure 3-1.

Section 1 describes the design issues for MEMS mirrors made in the MUMPS fabrication process. Section 2 describes measurements of the reliability and robustness of MUMPS mirrors, and section 3 offers conclusions on the usefulness of the MUMPS fabrication process for scanning mirrors.

Table 3-1. Simplified MUMPS process flow

Step	Process
1	Deposit silicon nitride, 0.6 microns
2	Deposit polysilicon layer (POLY0) , 0.5 microns
3	Pattern and etch POLY0, stop on nitride
4	Deposit oxide layer (OX1), 2 microns
5	Pattern and etch DIMPLE, 1 micron deep etch (no etch stop)
6	Pattern and etch ANCHOR1, stop on POLY0
7	Deposit polysilicon layer (POLY1), 2 microns
8	Pattern and etch POLY1, stop on OX1
9	Deposit oxide layer (OX2), 0.5 microns
10	Pattern and etch POLY1_POLY2_VIA, stop on POLY1
11	Pattern and etch ANCHOR2, stop on POLY0 or nitride
12	Deposit polysilicon layer (POLY2), 1.5 microns
13	Pattern and etch POLY2, stop on OX2
14	Pattern GOLD
15	Deposit gold, 0.5 microns
16	Lift off gold
17	Post processing (dicing, release, assembly, et cetera)

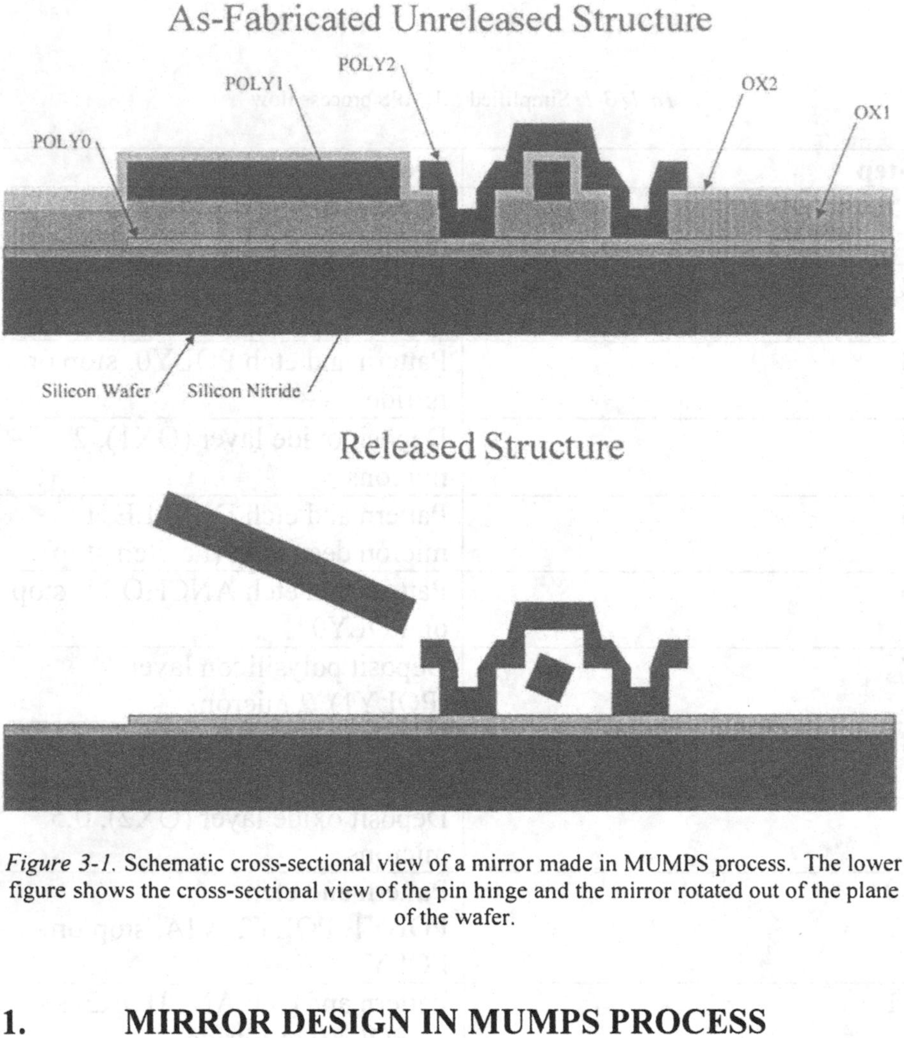
As-Fabricated Unreleased Structure
POLY2
POLY1
OX2
OX1
POLY0
Silicon Wafer
Silicon Nitride
Released Structure

Figure 3-1. Schematic cross-sectional view of a mirror made in MUMPS process. The lower figure shows the cross-sectional view of the pin hinge and the mirror rotated out of the plane of the wafer.

1. MIRROR DESIGN IN MUMPS PROCESS

The Multi-User MEMS Process (MUMPS) is a commercially available surface-micromachining fabrication process that can be used for both low-volume prototyping and higher-volume production. The process is derived from the surface-micromachining processes developed at the Berkeley Sensor & Actuator Center (BSAC) at the University of California Berkeley in the 1980's and early 1990's.

An overview of mirror design in the MUMPS process is given in section 1.1. A more detailed description of the hinges used for fabrication of MUMPS mirrors can be found in section 1.2. Section 1.3 describes static mirror flatness of MUMPS mirrors, and section 1.4 shows results from dynamic deformation measurements of these mirrors. A brief discussion of electrostatic combdrive actuator design in the MUMPS process is given in section 1.5.

1.1 Overview

Mirror designs made in the MUMPS surface-micromachining process are fold-up mirrors; they are fabricated in the planar process and are folded up away from the substrate after the final oxide etch releases the structures [33]. The original fold-up structures were made by Kris Pister [34], [35], and a good deal of further research has enabled its extension through the inclusion of off-substrate "finger" hinges, "locking" hinges, and others. With the two structural polysilicon layers available in the MUMPS process, it is possible to make a variety of fold-up mirror structures such as that shown in Figure 3-2.

A variety of micromirrors have been designed in the MUMPS micromachining process, including mirrors that lift up off the substrate [36], [37], mirrors that scan along an axis normal to the substrate [38], and mirrors that scan along an axis parallel to the substrate [39], [40]. These mirrors have been moved using thermal actuators [41], [42], [43], electrostatic gap-closing actuators [44], [45], [46], linear electrostatic combdrives [39], linear electrostatic vibromotors [47], and electromagnetic drives [48]. Our work has concentrated on linear electrostatic-combdrive-actuated micromirrors, similar to those originally described by Tien, Solgaard, Kiang, Daneman, Lau, and Muller [39], using actuators similar to those originally described by Bill Tang [49]. Many of our results and derivations are, however, applicable to a wider range of designs.

An electrostatically actuated scanning micromirror is shown in Figure 3-2. The comb-drive actuator, at the bottom of the figure, generates a force away from the mirror when a voltage is applied. This force is extended to the mirror by the coupling hinge as shown in the figure. The mirror is attached to the frame by two torsion hinges – formed from bars of polysilicon connected to the mirror. The torsion bars twist when a torque is applied to the mirror. The frame is supported at the bottom by substrate hinges, and at the back by the backframe. The backframe is attached to a slider that is constrained by a hub so that it slides parallel to the substrate. The mirror is fabricated in the plane of the wafer, and is folded up after release by pushing the slider towards the mirror with a microprobe. The mirror frame and backframe fold up out of the plane of the substrate as the slider moves towards the mirror, and when fully folded up the slider is held in place by friction between it and the substrate.

Both the theory of operation of these surface-micromachined mirrors, and a number of applications for them have been presented [7], [21], [54].

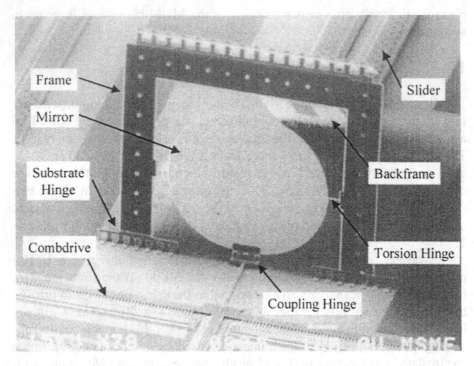

Figure 3-2. Fold-up mirror structure from MUMPS process.

Figure 3-3. SEM of MUMPS fold-up micromirror.

1.2   Mechanical linkages

The mechanical linkages, such as the coupling hinge that connects the actuator to the mirror, are an important aspect of the design because improper hinge design can add unwanted mirror motion and can reduce the mirror reliability. There are two categories for linkage functionality: fold-up linkages and coupling linkages. We will discuss the benefits and drawbacks of various fold-up and coupling linkage designs below.

1.2.1   Fold-up linkages

The fold-up linkages are used only during the mirror assembly process and should provide very rigid support after mirror assembly. During mirror operation, significant forces are applied to the fold-up linkages, and any play in these linkages can cause unwanted and uncontrollable hinge motion. Surface forces at these hinges are sometimes sufficient to prevent unwanted hinge motion, but for many mirror designs the forces exerted on these hinges can be large enough to cause the mirror frame to jump unpredictably from one state to another [27], a serious problem in applications where high-accuracy beam steering or scanning is necessary. This problem has been addressed with improved designs that reduce the play and backlash in the hinges [50].

Figure 3-4 shows an SEM of a mirror with glue at both the substrate hinge and the backframe hinge, which prevents unwanted and irregular mirror motion.

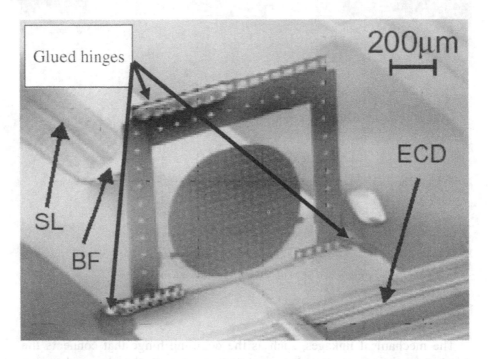

Figure 3-4. MUMPS mirror with glued substrate and frame hinges. SL indicates the slider, BF indicates the backframe, and ECD indicates the electrostatic combdrive.

The mirror and frame are fabricated in the plane of the substrate and are folded out of plane after release. For the mirror shown in Figure 3-2, the slider (which is constrained by a hub to only move linearly in the plane of the wafer) is pushed forwards towards the mirror frame and substrate hinge to lift the frame and backframe up off the plane of the wafer.

This assembly procedure is somewhat difficult and failure-prone; in a lab environment using a manual or automatic probe station a significant fraction of the mirrors are damaged during this assembly procedure. Work has been done to automate this fold-up procedure [51], [52], [53], but this assembly step will ultimately reduce yield and add cost to surface-micromachined MEMS mirrors.

In order to eliminate this costly post-fabrication assembly, we have fabricated a number of mirror frames that might allow fluidic self-assembly of the mirror frame upon release using designs similar to those previously described [54]. The fluidic self-assembly hinge consists of a long bar of polysilicon attached to the substrate at one end some distance from the frame substrate hinge, and a pair of inward-pointing triangles at the other end. The frame has a hole and a slot designed such that the bar will fit into the slot and firmly hold the mirror frame up from the substrate. This mechanism locks the mirror frame in place after the frame is rotated perpendicular to the substrate. During the liquid HF release process, the frame is often rotated perpendicular to the substrate by fluidic forces, assembling the mirror without any outside intervention. Figure 3-5 shows a schematic of the design of one of these locking mechanisms.

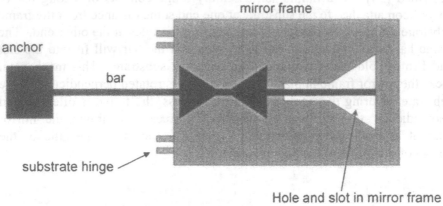

Top view (unassembled)

mirror frame

anchor

bar

substrate hinge

Hole and slot in mirror frame

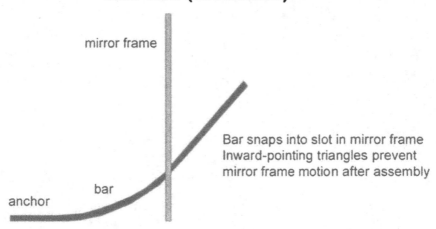

Side view (assembled)

mirror frame

Bar snaps into slot in mirror frame
Inward-pointing triangles prevent
mirror frame motion after assembly

bar

anchor

Figure 3-5. Schematic of fluidic self-assembly mirror frame structure with locking mechanism.

The structure was fabricated in the MUMPS process, and worked as expected. Figure 3-6 shows an SEM of a mirror frame with a locking mechanism used to hold the mirror frame up from the substrate, and Figure 3-7 shows a close-up SEM of the locking mechanism. This design has three major advantages over the slider and backframe approach: 1) assembly is potentially automatic from fluidic forces during release; 2) the locking mechanism can very accurately fix the mirror position without the hinge and slider backlash and motion that occurs in the slider-and-backframe design; and 3) the locking mechanism takes very little chip area compared to the slider-and-backframe design, thereby saving chip area and ultimately reducing mirror cost. We discovered no significant drawbacks of this design.

Figure 3-6. of surface-micromachined mirror with frame-locking mechanism.

Figure 3-7. Close-up SEM of frame locking mechanism.

1.2.2   Coupling linkages

The coupling linkages are in motion during device operation, and should ideally provide a direct coupling of the force from the actuator without exerting out-of-plane forces on the actuator, even while the mirror is moving. These linkages typically convert linear motion in the plane of the substrate to rotational motion of the mirror. Figure 3-8 shows an SEM of a finger hinge used to convert linear motion of an actuator to rotational motion of a mirror. The difficulties with this hinge are: 1) the hinge has at least ±1 μm of motion in the direction of actuation which cannot be controlled accurately, 2) the hinge is not flexible in the vertical direction, so as the mirror rotates about the hinge, the actuator pulls up from the plane of the substrate, 3) long-term operation of this hinge may cause wear at the interface resulting in particle generation that could short the combdrive or cause other reliability problems.

These complications can all be addressed by using compliant hinges – hinges that rely on the bending of mechanical elements rather than on surface contact. A variety of different compliant hinge designs have been fabricated, all of which address the three issues described above. Figure 3-9 shows a compliant hinge design that uses a long bar parallel to the mirror surface as the connection. This hinge is rotationally compliant, so as the frame is rotated up off the substrate the hinge bar twists, keeping the actuator in the plane of the wafer. The compliance in the direction of actuator motion, however, is nonlinear, so the actuator force is stiffly coupled to the mirror. This hinge does not exhibit surface-to-surface contact that could generate particles (which can ultimately reduce mirror lifetime), nor does the hinge have any irregular motion from hinge backlash.

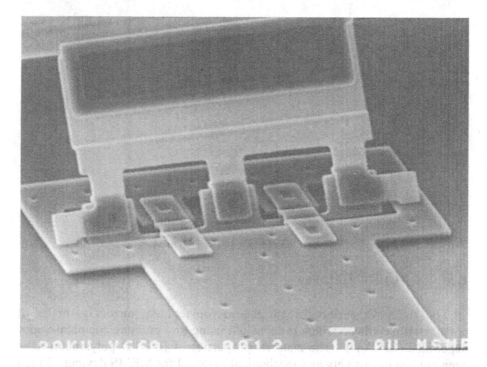

Figure 3-8. SEM of coupling hinge connecting electrostatic comb-drive actuator with mirror.

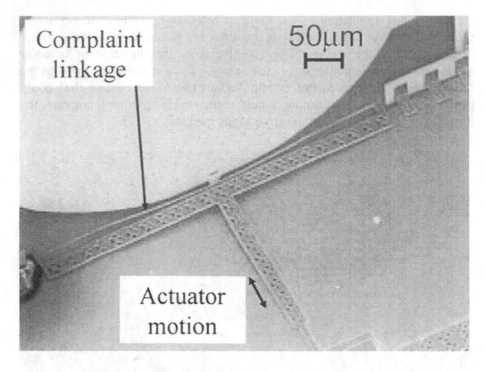

Figure 3-9. Compliant linkage hinge fabricated in MUMPS process

1.3   Mirror flatness

The reflective surface of surface-micromachined mirrors is made of polycrystalline silicon which is deposited using low-pressure chemical-vapor deposition (LPCVD). The material properties of LPCVD polysilicon have been studied extensively as a mechanical material for MEMS devices. Much of this work has focused on reducing the stress in the deposited material layer to reduce device curvature that can complicate designs [55], and surface-micromachining process modifications have been designed to reduce the effects of residual film stress [22].

For optical systems mirror flatness is paramount – deviations from planar on the order of $\lambda/8$ (82 nm for 655 nm light) can reduce the optical resolution (see Chapter 2 section 2 for a description of the effects of mirror deformation on optical performance). This places a stringent requirement on both the stress and the stress gradients in the deposited polysilicon used for successful fabrication of high-quality mirrors. While some recent results have shown good surface flatness control, previous MUMPS runs had peak-to-trough surface deformations of more than 1 μm for 500 μm-diameter mirrors. Figure 3-10 shows the surface-height map of the mirror shown in

Figure 3-4 in the rest position (static). The figure shows more than 1.6 μm of static nonplanarity.

While this static deformation can be partially corrected using lenses or other optical elements, the variation from mirror to mirror (due to slight fabrication process variations) makes this tedious and expensive. In addition, the surface deformation measured on MUMPS mirrors is not spherical – Figure 3-11 shows a surface height map from the same mirror with the defocus term of the optical aberration removed – so custom optics would be required to achieve diffraction-limited optical performance.

The mirrors presented here are all made of polysilicon without a reflective coating. The addition of a reflective coating can add significantly to the static nonplanarity of the mirrors. Even if the mirrors with the reflective coating were flat, significant static mirror curvature could result from the differing coefficients of thermal expansion of the polysilicon and the reflective coating. Thicker mirrors (where the ratio of the mirror thickness to the reflective-coating thickness is larger) exhibit less curvature due to the bimorph effect than the 1.5-3 μm-thick surface-micromachined mirrors [19], [56].

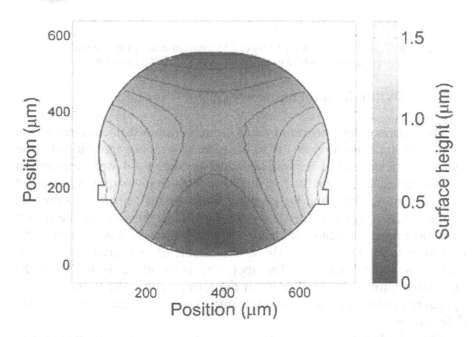

Figure 3-10. Static deformation (in μm) of a 550 micron-diameter mirror made in the MUMPS process showing more than 1.5 microns of peak-to-trough nonplanarity (figure courtesy of Matthew Hart [27]).

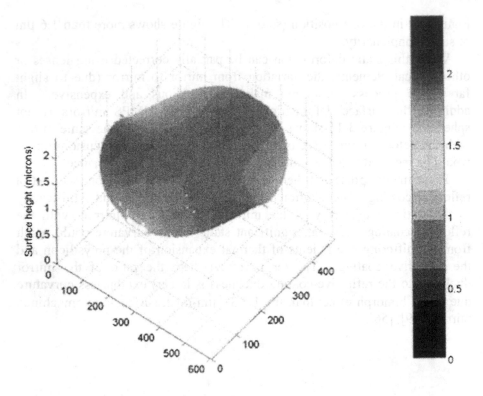

Figure 3-11. Static deformation (in μm) of a 550 micron-diameter mirror with focus correction showing total nonplanarity greater than 200 nanometers.

1.4   Dynamic deformation

The dynamic deformation of surface-micromachined mirrors presents a significant limitation to their use in high-speed scanning applications. The dynamic deformation of the mirror shown in Figure 3-4 was measured at various times during one half of its period while moving at its resonant frequency of 4.4 kHz, and the captured surface-height maps are shown in Figure 3-12a. Figure 3-12b shows the surface-height maps of the mirror with tilt removed, which more clearly shows the dynamic deformation of the optical surface. Figure 3-12c shows the calculated and measured far-field spot sizes at each instant. The spot size is clearly enlarged from the diffraction-limited spot size shown at the lower right of the figure, and the optical resolution of the scan is significantly degraded by the mirror dynamic deformation.

Not only does the mirror show significant variation in deformation over the scan, but the frame to which the mirror is attached can also have unacceptable levels of dynamic deformation that can contribute to motion in other vibratory modes. Figure 3-13a shows the magnitude of the motion at

104 points on the mirror surface versus frequency, and Figure 3-13b shows the line over which the 104 points were taken. The groove in the response shown in Figure 3-13a (the low-amplitude response at low-frequency) shows that, as expected, there is little or no vertical motion at the axis-of-rotation (at a distance of 200 microns up from the bottom of the mirror). Above the primary resonance at 4.43 kHz there are two other significant modes of vibration, both of which have significant motion at the axis-of-rotation. Figure 3-13d and e show surface contours of these two higher-order modes. The mode at 8.03 kHz shown in Figure 3-13d corresponds to the mirror rotating along an axis perpendicular to the axis-of-rotation of the 4.43 kHz mode (known as mirror wobble), which is likely due to bending of the mirror frame. The vibratory mode shown in Figure 3-13e corresponds to bending of the mirror and frame at 20.61 kHz.

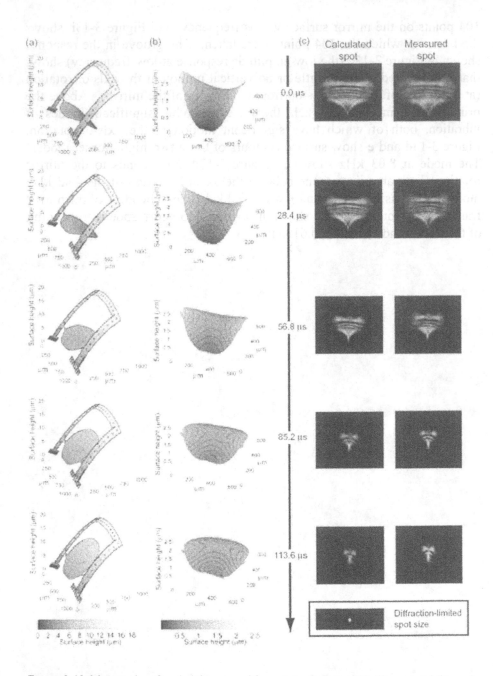

Figure 3-12. Measured surface-height maps of frame (a), of mirror with tilt removed (b), and far-field spot size calculated from mirror surface-height measurement (c) and measured with CCD camera (d) for surface-micromachined mirror scanning at 4.4 kHz (figure courtesy of Matthew Hart [27]).

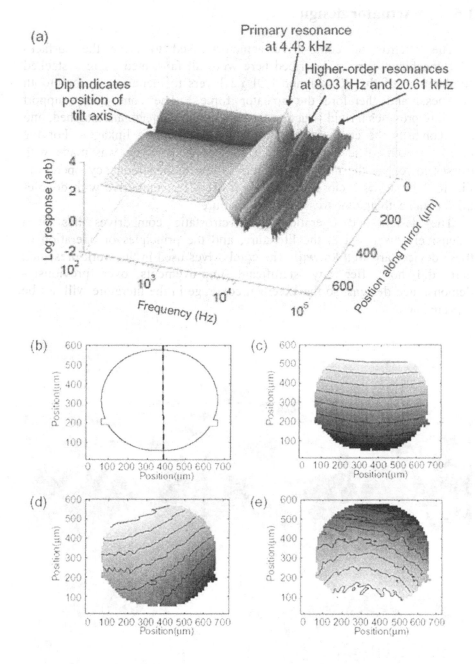

Figure 3-13. a) dynamic response of mirror and frame over various frequencies; b) line showing 104 points over which data was analyzed for a); c) first mode of vibration showing mirror tilting at 4.43 kHz; d) second mode of vibration showing mirror wobble at 8.03 kHz; e) third mode of vibration showing mirror bending at 20.61 kHz. (figure courtesy of Matthew Hart [27]).

1.5   Actuator design

The electrostatic comb-drive actuators used to drive the surface-micromachined mirrors described here were all fabricated using a stacked combination of the POLY1 and POLY2 layers to increase the combtooth thickness, and therefore the actuator force. The comb-drive support structure provides a rigid structure to which the combteeth are attached, and also connects the comb-drive actuator to the coupling linkage. For the mirrors presented here, the comb-drive support structure was made with trusses to reduce the mass and thereby allow higher-frequency operation. Figure 3-14 shows a close-up SEM of a MUMPS combdrive with double-thick combteeth and the truss support structure.

The design and operation of electrostatic combdrives has been extensively discussed in the literature, and the principles of operation of these devices are well known. The combdrives used in the work presented here did not offer any significant improvements over previously-demonstrated designs, so the excellent coverage in the literature will not be repeated here.

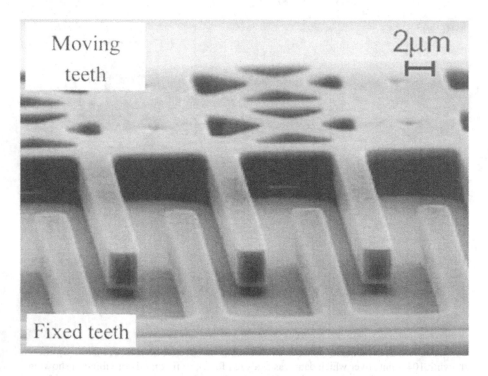

Figure 3-14. MUMPS electrostatic comb-drive actuator.

2.   ROBUSTNESS AND RELIABILITY

The perceived fragility of structures that are 1-5 µm-thick and nearly 1 mm tall, like the scanning micromirror shown in Figure 3-2, have led to questions about the reliability and robustness of such structures. To explore the validity of these concerns, we performed a series of tests on scanning micromirrors. The robustness and reliability was characterized by measuring time-to-failure, positioning repeatability, and temperature sensitivity, and by performing shock testing.

2.1   Mirror lifetime

Long-term reliability is a critical issue for any commercially viable system. Conventional scanning mirrors use mechanical bearings, which are subject to wear and eventual failure. Many micromachined structures also fail due to wear of their contacting surfaces [57]. Micromachined mirrors, however, can be designed as completely compliant structures with no contacting surfaces, thereby eliminating the inter-surface frictional wear that limits the lifetime of some conventional optical scanning mirrors. The mirror shown in Figure 3-2 has only one noncompliant element in the drive train – the hinge connecting the combdrive to the mirror. This hinge should have little or no motion after mirror assembly, which reduces the likelihood of hinge failure due to surface wear and combdrive failure due to particle generation at the hinge.

Another possible mode of failure for surface-micromachined scanning mirrors is material fatigue. Material fatigue is most likely to occur at the point of highest strain in the structure, which is at the torsion hinge that attaches the mirror to the frame. At maximum deflection (10°), the strain in the torsion hinge is approximately 0.6% – well below the yield strain of polysilicon. Numerous mechanisms for fatigue of polysilicon have been proposed [58], but the mechanisms are still poorly understood, so we have empirically characterized the lifetime of one scanning micromirror by operating it at its 5.4 kHz resonant frequency for an extended period of time. By measuring the resonant frequency periodically throughout the 97-day duration of the test, we were able to detect small changes in the mirror behavior that could be indicative of material fatigue. A typical frequency response of the scanning micromirror used in this test is shown in Figure 3-15 and the results from the test are shown in Figure 3-16. We found that the change in resonant frequency over the 45 billion cycles was less than 1.5% of the resonant frequency, and the trend was not monotonic. This indicates that the dominant variation may be due to temperature or humidity changes rather than material fatigue, since fatigue and crack growth would be expected to cause a monotonic decrease in the resonant frequency.

After 45 billion cycles, the electrostatic comb-drive fingers became electrically shorted. Visual inspection showed this to be a result of contamination from the open-air lab environment. Proper packaging can eliminate this mode of failure, potentially extending the mirror lifetime beyond 45 billion cycles.

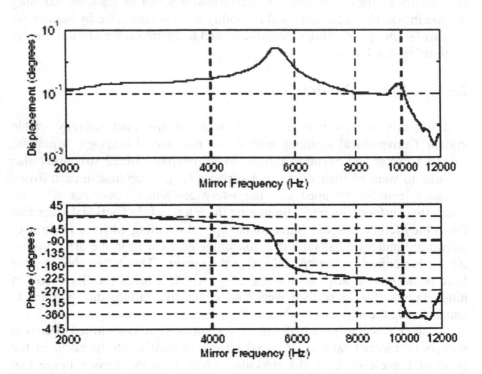

Figure 3-15. Magnitude and phase of scanning micromirror versus drive frequency (mirror motion is at twice the drive frequency since the force of the comb-drive actuator is proportional to the square of the applied voltage).

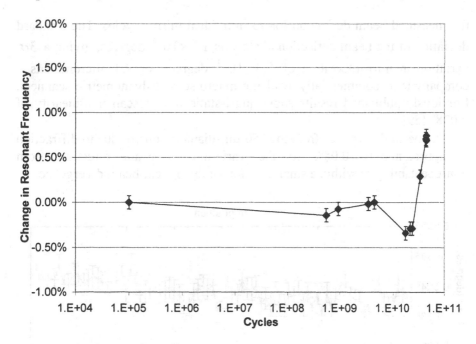

Figure 3-16. Measured resonant frequency over 45 billion cycles of line-scan mirror.

2.2 Positioning repeatability

Mirror positioning and resonant mirror-scan repeatability are also important for scanning micromirrors. The scan and positioning repeatability should ideally be comparable to, or better than, conventional macro-scale galvanometric scanners (typically 1-10 microradians). For a 550 μm-diameter mirror with a torsion hinge attached at the centerline of the mirror, 10 microradian angular-positioning accuracy corresponds to 2.75 nm positioning accuracy at the joint between the combdrive and the mirror.

By observing the micromirror structure under an optical microscope, we were able to see that the polysilicon pin hinges that attach the mirror to the frame and the scissor hinges that connect the elements of the frame move intermittently while the mirror was scanning, thereby seriously degrading the positioning repeatability of the micromirror. Using UV-curable epoxy, we secured both of the frame hinges in order to prevent frame motion. The only remaining free hinges are the torsion hinge attaching the mirror to the frame (which is a compliant hinge, and therefore unlikely to cause a problem with repeatability) and the hinge connecting the combdrive to the mirror.

After securing the frame with UV-curable epoxy, we characterized the open-loop positioning repeatability of the micromirror by driving the mirror with a square wave and measuring the mirror response. Figure 3-17 shows the measured beam deflection for 75 individual mirror cycles. The standard deviation of the beam deflection angle was 1.8×10^{-4} degrees, giving a 3σ variation in mirror position of 5.4×10^{-4} degrees, or 9.4 microradians – comparable to commercially available macro-scale galvanometric scanners. Previously published results show quasi-static mirror scan nonlinearity of 0.038° [59].

The beam divergence from the 550 μm-diameter mirror due to diffraction is approximately 0.08°, so the variation in beam direction from nonrepeatability is within a small fraction of the optical-beam divergence.

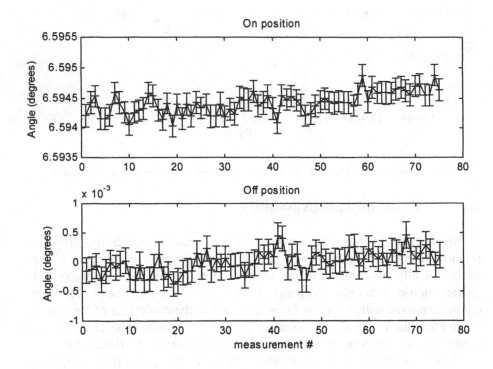

Figure 3-17. Consecutive position measurements from square wave drive signal, with error bars indicating standard deviation of the mean calculated from detector electrical noise.

2.3   Thermal effects

Commercial or industrial use of scanning micromirrors will require that the mirror characteristics, in particular the resonant frequency and steady-state mirror position, be fairly stable with changes in temperature. Conventional commercial closed-loop galvanometric scanners have temperature sensitivities of 100-200 ppm/°C, and resonant scanners typically have resonant-frequency temperature sensitivities on the order of 80 ppm/°C.

2.3.1   Theory

The dependence of the resonant frequency on temperature can be predicted from the coefficient of thermal expansion and the temperature coefficient of the Young's modulus. The folded-spring stiffness k for the surface-micromachined mirrors described here is a combination of the torsional stiffness of the torsion beam which connects the mirror to the frame and the linear stiffness of the electrostatic comb-drive suspension. The total spring stiffness can be represented as a torsional stiffness about the axis-of-rotation, or equivalently as a linear stiffness in the plane of the actuator. The linear stiffness k_{lin} is proportional to the various hinge linear dimensions (which we will denote by the variable L) and Young's modulus E as

$$k_{lin} \propto EL \qquad [3\text{-}1]$$

For an isotropic material, all of the length terms have the same temperature dependence, so for a change in temperature ΔT, the stiffness changes from the original value according to the relation

$$k_{lin}(T) = k_{lin}(T_{ref})(1 + c_E \Delta T)(1 + \alpha \Delta T) \qquad [3\text{-}2]$$

where α is the coefficient of thermal expansion and c_E is the temperature coefficient of the Young's modulus. If the mass M of the system stays the same with temperature, then the resonant frequency ω scales with temperature as

$$\omega(T) = \sqrt{\frac{k_{lin}(T)}{M}} = \omega(T_{ref})\sqrt{(1 + c_E \Delta T)(1 + \alpha \Delta T)} \approx \omega(T_{ref})\left[1 + \frac{(c_E + \alpha)}{2}\Delta T\right]$$

$$[3\text{-}3]$$

The approximation in Equation [3-3] is valid if $c_E \Delta T, \alpha \Delta T \ll 1$. The coefficient of thermal expansion for bulk silicon over the range of temperatures from 300 to 1500K is

$$\alpha = 3.725 x 10^{-6}\left[1 - e^{-5.88 x 10^{-3}(T-124)}\right] + 5.548 x 10^{-10} T \qquad [3\text{-}4]$$

where T is the temperature in Kelvin [60]. Values reported for polysilicon at 300K are similar (2.9×10^{-6} /° C [61]), but is likely to be dependent on the deposition conditions and the grain structure of the polysilicon. The temperature coefficient of the Young's modulus for bulk crystalline silicon for temperatures between 300 and 1000K is 9.4×10^{-5} /° C [62]. Using the material properties of bulk silicon, the temperature sensitivity of the resonant frequency at room temperature (according to Equation (3-3)) is –46 ppm/°C (nearly linear for small variations).

The change in resonant frequency can be related to the change in steady-state deflection of the micromirror if we assume that the actuator force does not change with temperature. The mirror position θ is given by

$$\sin(\theta) = \frac{FH}{k_{lin}(T)} \tag{3-5}$$

where F is the force from the comb-drive actuator and H is the distance from the axis-of-rotation of the mirror to the connection with the actuator, as shown in Figure 3-18. Using Equation (3-5) and recognizing that H also changes with temperature, the steady-state deflection is

$$\sin(\theta) = \frac{\sin(\theta_0)}{(1 + c_E \Delta T)} \approx \sin(\theta_0)(1 - c_E \Delta T) \tag{3-6}$$

The scale factor is approximately +94 ppm/°C for silicon.

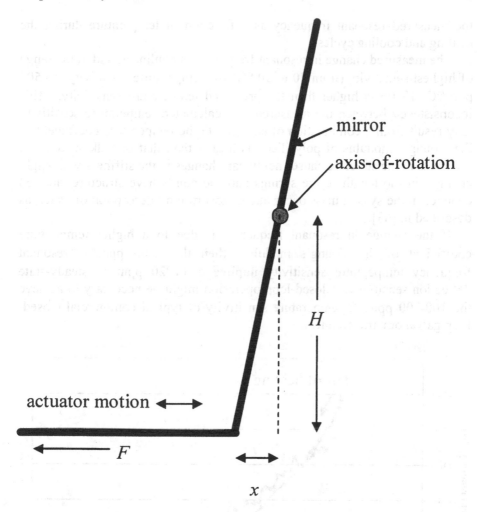

Figure 3-18. Schematic side view of surface-micromachined mirror.

2.3.2 Measurements

We measured the resonant frequency of the scanning micromirror using a spatially filtered beam from a 635 nm laser diode reflected from the micromirror onto a position sensitive diode (PSD). A lock-in amplifier (Stanford Research SR850) scanned the frequency of the micromirror drive signal and measured the corresponding magnitude and phase of the PSD output at the appropriate frequency. With the scanning mirror mounted on a hotplate, we measured its resonant frequency and temperature as the hotplate heated to 100 °C and then cooled to room temperature. Figure 3-19 shows the measured resonant frequency as a function of temperature during the heating and cooling cycles.

The measured change in resonant frequency is nonlinear, and in the range of highest sensitivity (from 20 to 50 °C) the temperature sensitivity is −500 ppm/°C, 10 times higher than the predicted temperature sensitivity. This inconsistency between the measured and calculated temperature sensitivity may result from a combination of causes: (1) the temperature coefficient of the Young's modulus of polysilicon is higher than that of bulk silicon, (2) the stress in the structure causes nonlinear changes in the stiffness with slight changes in the length of the springs and the comb-drive structure, and (3) changes in the system mass occur due to absorption / desorption of water, as described in [63].

If the change in resonant frequency is due to a higher temperature coefficient of the Young's modulus, then the −500 ppm/°C resonant frequency temperature sensitivity implies a +1020 ppm/°C steady-state deflection sensitivity. Closed-loop operation might be necessary to achieve the 100-200 ppm/°C temperature sensitivity of typical commercial closed-loop galvanometric scanners.

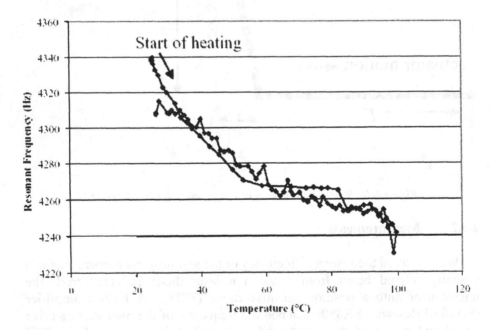

Figure 3-19. Change in mirror resonant frequency with temperature.

2.3.3 Mechanical shock testing

Compared to conventional optics, MEMS scanning micromirrors look extremely fragile. However, because the masses of the structures are miniscule, the mirrors are quite resistant to shock and vibration. When acted upon by forces from the macro world (such as a finger or a probe tip) the structures break fairly easily, but the forces exerted by acceleration scale as the length cubed. The strength scales as the length squared, so as the length of a device shrinks, it becomes sturdier.

The mass of the combdrive and scanning-mirror assembly is estimated to be approximately 1.6 µg, so to exert a 1 µN force (approximately equivalent to the force exerted by the electrostatic combdrive), the acceleration must be more than 60 times the acceleration of gravity. In order to verify the robustness of the micromirrors, we packaged a mirror in a metal box and performed the drop test specified by the ANSI/ISA standard [64]. The mirrors repeatedly survived the drop test without measurable changes in resonant frequency and without any visible damage.

3. CONCLUSIONS ABOUT SURFACE-MICROMACHINED MIRRORS

Surface-micromachined mirrors have some compelling features: they are easy to manufacture using planar semiconductor manufacturing processes; they may be compatible with CMOS circuit fabrication; and they are mechanically robust and reliable.

However, there are also some significant disadvantages to surface-micromachined mirrors. Surface-micromachining fabrication processes are typically limited to mirrors thinner than 10 microns, which severely limits the range of operating frequencies over which they are useful for high-resolution resonant-scanning applications. Mirror curvature due to stress gradients in the thin-film depositions presents a challenge for process designers attempting to make flat mirrors. Post-fabrication assembly poses formidable yield issues, so large-scale application of surface-micromachined mirrors will require technology development for assembly techniques. The thin mirrors are also subject to temperature-dependent static deformation from the bimorph effect which can reduce optical performance.

Ultimately, surface-micromachined mirrors are not suitable for low-cost, high-speed resonant scanners, but they may prove applicable for certain low-speed static-beam-steering applications.

Chapter 4

Staggered Torsional Electrostatic Combdrive (STEC) Micromirror

High-aspect-ratio micromachining processes have been developed over the past few years, especially enabled by new deep-trench silicon etchers that use the Bosch process (65]. This process enables etching of trenches in silicon with aspect ratios up to 100:1. This deep-trench-etching process provides a hybrid between surface micromachining, as described in Chapter 3, and traditional bulk micromachining, where wet chemical etching is used to make structures from single-crystal silicon. High-aspect ratio micromachining using deep-trench silicon etching has been applied to a number of MEMS mirror designs. The work we present in this chapter shows that high-aspect-ratio micromachined mirrors can deliver significant performance improvements over surface-micromachined mirrors because: a) the thicker mirrors reduce dynamic deformation; b) the process allows designs without hinges that reduce the accuracy and repeatability of the mirror scan; and c) combtooth thicknesses are increased to provide higher-torque actuators. Section 1 describes the Staggered Torsional Electrostatic Combdrive (STEC) micromirror used to make high-speed, high-resolution scanning mirrors, and section 2 describes the Tensile Optical Surface (TOS) process modification that allows lighter-weight mirrors for higher-speed, steady-state beam steering.

1.   STEC MICROMIRRORS

1.1   Overview

Scanning mirrors have many potential applications: barcode readers, laser printers, and confocal microscopes are just a few possible applications. Most of the scanning mirrors commercially available today are macro-scale galvanometers or polygonal rotating mirrors, which suffer from significant performance limitations related to scanning speed, power requirement, cost, and size. These limitations often make them unsuitable for high-speed and portable systems. Faster and smaller scanning mirrors could enable new applications such as raster-scanning projection video displays [12], [66], fiber-optic attenuators [67], and fiber-optic switches [68], [69], [70], [71], [72], [73]. They could also significantly improve performance in existing applications such as laser printers. Microelectromechanical Systems (MEMS) technologies enable manufacture of higher-speed scanners with lower power consumption and potentially lower costs, thereby making these new applications commercially viable.

A wide variety of MEMS scanning mirrors have been presented using an assortment of fabrication processes, including surface-micromachining as described in Chapter 3 and [59], [74], bulk micromachining [1], [75], [76], [77], and high-aspect ratio approaches [78], [79], [80], [81], [82]. Thus far, none has demonstrated the ability to simultaneously meet both the high scan speed *and* high resolution desirable for applications such as raster-scanning projection-video displays, two-dimensional raster-scanning barcode readers, and high-speed laser printers. For example, surface-micromachined scanning mirrors driven by electrostatic combdrives have been shown to operate at high scan speeds (up to 21 kHz). In these mirrors, however, static and dynamic mirror deformations limit resolution to less than 20% of the diffraction limit [27]. Magnetically actuated mirrors have been demonstrated with high speeds and large amplitudes [83], but they have not yet demonstrated high resolution, and they often require off-chip actuation which can limit the cost advantages promised by MEMS technologies.

In order to take advantage of the unique capabilities of micromachining technology, we have fabricated a new micromirror and shown it to be capable of high-speed, high-resolution scanning with low power consumption. This micromirror, the Staggered Torsional Electrostatic Combdrive (STEC) micromirror, has the low dynamic deformation characteristic of bulk-micromachined mirrors and the high-efficiency electrostatic comb-drive actuator often used with surface micromachined mirrors.

Figure 4-1 is a schematic view of the Staggered Torsional Electrostatic Combdrive (STEC) micromirror, which is made from two layers of single-crystal silicon separated by a 1.7 µm-thick silicon dioxide layer. The fabrication process is described in section 1.2.

The mirror, torsion hinge, and moving combteeth are in the top silicon layer; fixed combteeth are in the bottom silicon layer (the substrate). Applied voltage between the top and the bottom layers attracts the moving combteeth to the fixed combteeth, exerting torque on the mirror and causing it to tilt. Mechanical strain in the torsion hinges, which are anchored to the bottom silicon layer, provide restoring torque. Similar torsional actuators have been described earlier [84], [85]. A more detailed description of the actuator theory of operation is given in section 1.3.

The measured results from experiments that characterize the STEC mirror scan speed, optical resolution, and power consumption are described in section 1.4.

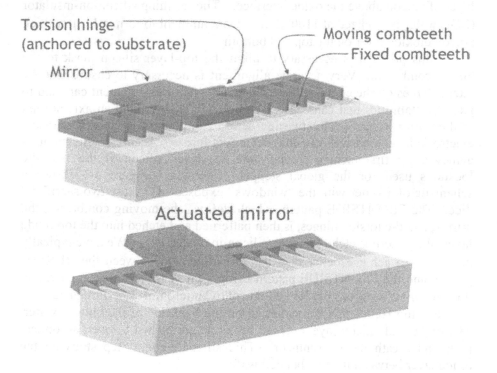

Figure 4-1. Schematic view of STEC micromirror with actuator.

1.2   Fabrication process

The STEC micromirrors are fabricated using deep reactive-ion etching and a bond-and-etch-back process with a buried pattern. The three-mask process flow is shown in Figure 4-2, and a detailed step-by-step description of the process flow is given in Appendix A

First, a silicon wafer is oxidized in steam at 1000°C to grow 0.2 μm of thermal oxide. This wafer is patterned with the BURIED pattern, and 100 μm-deep trenches are etched into it using an STS® deep-reactive-ion etcher to form the fixed combteeth. This wafer, along with another wafer having 1.5 μm of thermal oxide, is then cleaned and the two wafers are bonded together with the fixed combteeth at the wafer interface. The bonded wafer pair is then annealed at 1100°C for one hour to increase the bond strength [86].

Second, the bonded wafer is ground and polished to leave a 50 μm-thick layer of silicon above the oxide interface. The resulting silicon-on-insulator (SOI) wafer is oxidized at 1100°C in a steam ambient to form a 1.1 μm-thick silicon dioxide layer on the top and bottom.

At this point, it is necessary to align the top-layer silicon mask to the buried combteeth. Very accurate alignment is necessary to ensure that the lateral forces on the combteeth are balanced – any misalignment can lead to lateral instabilities that can cause the mirror to twist around an axis normal to the mirror surface. This twisting action can electrically short-circuit the electrostatic combdrive, causing device failure. Accurate alignment is achieved by first etching through the top-layer silicon at the two die locations used for the global stepper alignment. Figure 4-3 shows a schematic of a wafer with the "windows" exposed over these two sacrificial dice. The FRONTSIDE pattern, which defines the moving combteeth, the mirror, and the torsion hinges, is then patterned and etched into the top oxide layer (the pattern is etched into the silicon in a later step). We have typically been able to achieve better than 0.2 μm alignment between the BURIED pattern and the FRONTSIDE pattern using a GCA i-line wafer stepper, and this alignment error has not limited the STEC micromirror performance.

Next, the HOLE layer is patterned on the backside of the bottom wafer. The oxide and silicon layers on the backside are etched to open an optical path underneath the micromirror. This silicon-etching step stops on the oxide layer between the two bonded wafers.

The silicon on the top layer is then etched using the previously patterned top oxide layer as an etch mask. The structure is released with a timed HF etch to remove the sacrificial oxide film below the combteeth and mirror. Finally, a 100 nm-thick aluminum film is evaporated onto the optical surface of the mirror (either the top or the bottom, depending on the application) to increase its reflectivity for visible light to approximately 92% [87].

A scanning-electron micrograph (SEM) of a STEC micromirror chip is shown in Figure 4-4 with a close-up SEM of the electrostatic-comb-drive actuator, and Figure 4-5, Figure 4-6, and Figure 4-7 show other STEC mirror designs.

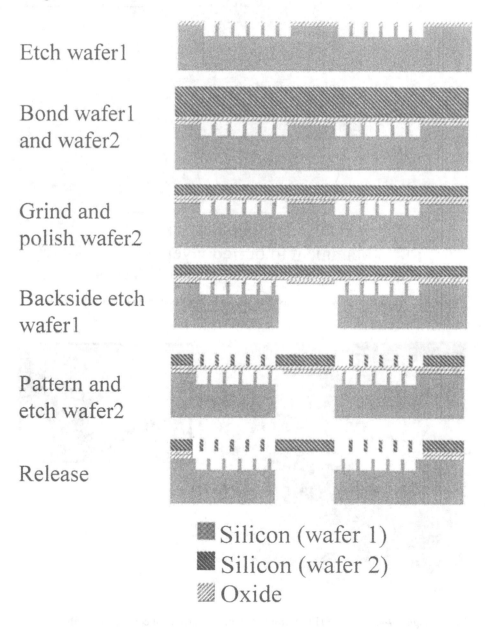

Etch wafer1

Bond wafer1 and wafer2

Grind and polish wafer2

Backside etch wafer1

Pattern and etch wafer2

Release

▨ Silicon (wafer 1)
▧ Silicon (wafer 2)
▨ Oxide

Figure 4-2. STEC fabrication process flow.

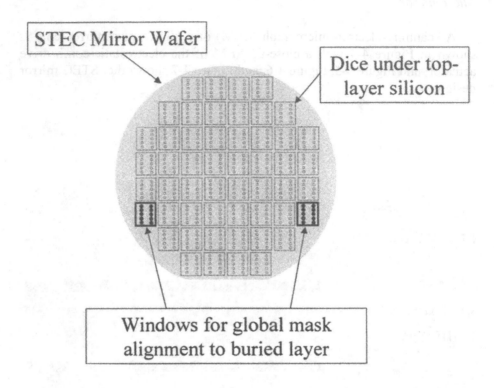

Figure 4-3. Schematic of "windows" used to align top-layer pattern to buried-layer pattern.

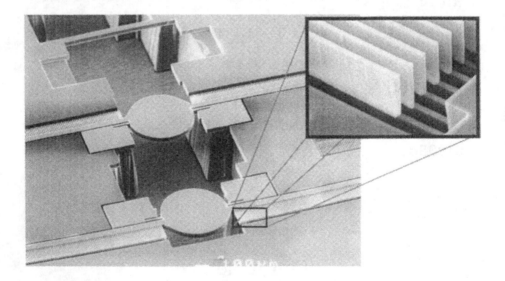

Figure 4-4. SEM of STEC mirror with close-up picture of comb-drive teeth.

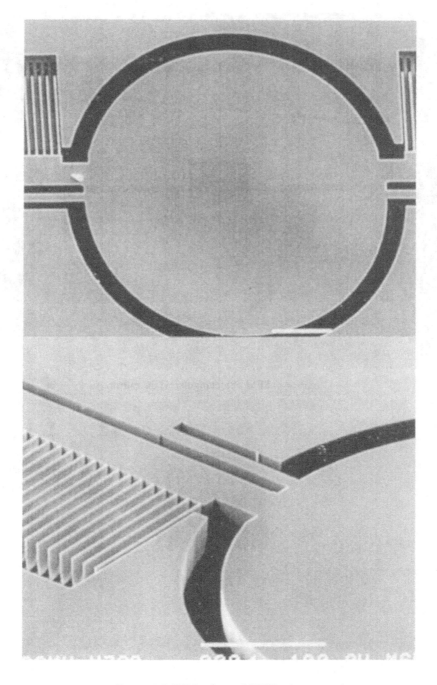

Figure 4-5. SEMs of round STEC mirror.

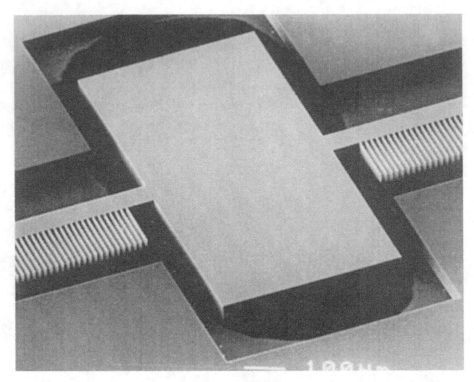

Figure 4-6. SEM of rectangular STEC mirror.

Figure 4-7. Close-up SEM of hinge at end of rectangular STEC mirror.

1.3   Actuator theory of operation

The torsional electrostatic combdrive of the STEC micromirror offers several advantages over previous electrostatic actuator designs. First, the STEC actuator generates a torque on the mirror directly rather than requiring a linkage to convert a linear motion to an angular motion, as was done in earlier surface-micromachined scanning mirrors using in-plane electrostatic [88] or thermal actuators [42]. These linkages complicate the dynamic behavior of scanning micromirrors and can introduce unpredictability of motion due to backlash and play in the hinges as described in Chapter 3 section 1.2 and [89].

Second, the STEC micromirror structure does not require post-fabrication assembly as is needed in the earlier surface-micromachined structures as described in section Chapter 3 section 1.2 and [52]. These post-fabrication assembly steps, which are typically carried out on an individual basis rather than as a batch process, reduce the manufacturing yield and add to the overall cost of manufacture. As mentioned in Chapter 3 section 1.2, novel assembly techniques [51, 53] may bring down the cost of this post-fabrication assembly, reducing this additional cost, but the STEC micromirror structure offers an attractive alternative to the cost and complexity of surface-micromachined mirror assembly.

Third, the moving and fixed combteeth of the STEC actuator are offset in the rest position, so it can be used for steady-state beam steering as well as for resonant scanning. Previously demonstrated balanced torsional electrostatic actuators (for which the moving and fixed combteeth were coplanar in the rest position) were suitable for resonant operation, but not for steady-state beam steering [90], [91].

Fourth, the torsional combdrive offers an advantage over gap-closing actuators because the energy density in the combdrive is higher than that in a gap-closing actuator, thereby allowing larger scan angles at high resonant frequencies with lower applied voltages.

Design of the STEC micromirror begins by choosing the desired optical resolution and resonant frequency, and deriving from these performance requirements the appropriate mirror size, as well as actuator and hinge geometries.

1.3.1   Optical resolution

The optical resolution of a micromirror is described in Chapter 2 section 2 above. The equations derived therein accurately describe the optical performance of STEC mirrors, including the effects of dynamic deformation. The mirror length can be determined from the necessary optical resolution using Equation (2-59).

1.3.2   Resonant frequency

For a high-Q system like the STEC mirrors presented here, the resonant frequency ω_r of the mirror is related to the moment-of-inertia I and the torsion-hinge stiffness k_θ as shown in Equation (2-5). The moment-of-inertia is given by Equation (2-22).

For the STEC mirror designs we discuss, the moment-of-inertia of the mirror is significantly larger than the moment-of-inertia of the actuator, so, to first order $\eta_{actuator} = 0$. The stiffness of the pair of rectangular torsion hinges is given in Equation (2-15), and the stress is given in Equation (2-16). These equations can be used to design a scanning mirror for a given resonant frequency ω_r.

1.3.3 Micromirror Design

With the equations above, the desired resonant frequency, optical resolution, and maximum numeric aperture (NA) can be used to determine the appropriate mirror length. The mirror width is often determined by limitations on the off-axis beam divergence, which affects the optical resolution perpendicular to the scan in the same way that the length affects the optical resolution along the direction of the scan as described in Equation (2-59).

The applications of scanning mirrors can be divided into two broad categories: steady-state beam steering and resonant scanning. Steady-state beam steering is used in applications such as optical switching where the purpose of the mirror is to reflect a beam of light in a particular direction, whereas resonant scanning is used to generate a sinusoidal mirror motion, such as is used in barcode scanning, laser printing, and video displays.

The design criteria for these two application classes are quite different, and are treated separately below.

1.3.3.1 Steady-state beam steering

For steady-state beam steering of the STEC mirror, the maximum achievable steady-state scan angle is related to the actuator torque T, which is

$$T = \frac{1}{2}V^2 \frac{dC}{d\theta} \tag{4-1}$$

where V is the applied voltage, and $\dfrac{dC}{d\theta}$ is the change in actuator capacitance with angle. A derivation of the electrostatic torque of a torsional combdrive is given in [92]. If the combtooth thickness is large compared to the electrostatic gap, a useful approximation for the actuator torque can be calculated by ignoring the fringing fields. In this case the capacitance between the fixed and moving combteeth is proportional to the overlap area $A(\theta)$ at the angle θ

$$C = \frac{2N\varepsilon_0}{g} A(\theta) \tag{4-2}$$

where ε_0 is the permittivity of free space, N is the number of moving combteeth, and g is the gap between the fixed and moving combteeth. The change in overlap area with angle is a nonlinear function of the mirror angle, increasing from the rest position until $\theta = \theta_0$ where the top of the moving combteeth overlaps with the top of the fixed combteeth (see Figure 4-8). At $\theta = \theta_0$, the torque is approximately

$$T_{max} \approx \frac{V_{max}^2 N\varepsilon_0\left(L_c^2 - d^2\right)}{2g} \tag{4-3}$$

where L_c is the distance from the axis-of-rotation to the tip of the combteeth and d is the distance from the axis-of-rotation to the start of the fixed combteeth (see Figure 4-8a). The angle at maximum torque is

$$\theta_0 = \sin^{-1}\left(\frac{t_c + g_0}{L_c}\right) \tag{4-4}$$

where t_c is the thickness of the combteeth, and g_0 is the gap between the top and bottom silicon layers. In Figure 4-9 we plot values for $\dfrac{dC}{d\theta}$ calculated for three representative combtooth lengths. For steady-state beam steering the scan angle for a given applied voltage is the solution of the transcendental equation

$$\theta(V) = \frac{T(\theta, V)}{k_\theta} = \frac{V^2}{2k_\theta}\frac{dC(\theta)}{d\theta} \tag{4-5}$$

Figure 4-9 shows the calculated values of $\dfrac{dC}{d\theta}$ for various combtooth lengths. The solution to Equation (4-5) corresponds to a line from the origin on Figure 4-9, the slope of which is equal to $\dfrac{2k_\theta}{V^2}$. The deflection angle for a particular applied voltage is the angle at the intersection of this line and the $\dfrac{dC}{d\theta}$ curve.

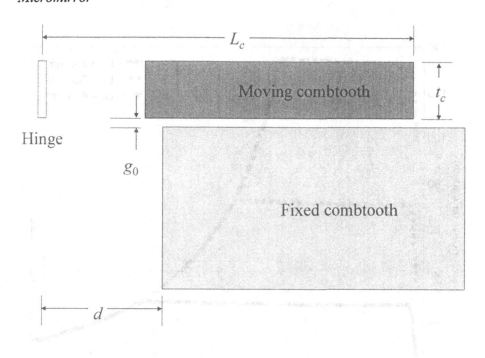

a) Rest position (no voltage applied)

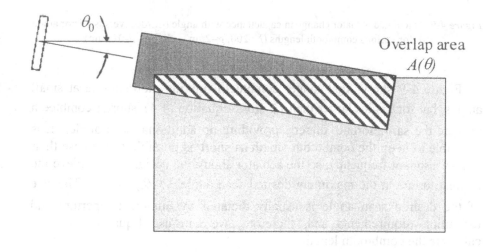

b) Offset position at which torque is maximum

Figure 4-8. Combtooth overlap with angle.

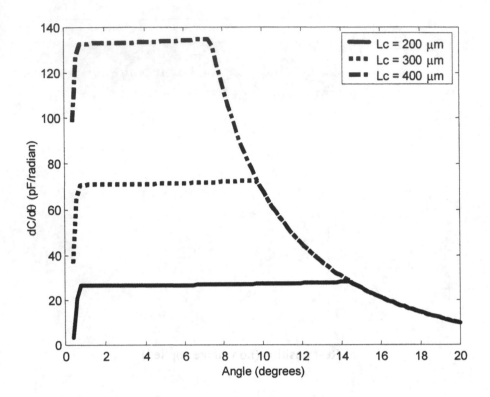

Figure 4-9. Calculated actuator change in capacitance with angle (dC/dθ) versus mirror angle (θ) for various combtooth lengths (N=200, g=2μm, t=50 μm, d=100μm).

Figure 4-9 shows that longer combteeth provide more torque at small angles, but for angles larger than θ_0 from Equation (4-4), shorter combteeth provide the same torque, thereby providing no additional scan angle. It is desirable to keep the combtooth length as short as possible to increase their lateral resonant frequency, so the actuator should be designed to achieve the highest torque at the maximum desired scan angle, so $\theta_{max} = \theta_0$. The size of the desired scan angle is usually dictated by numerical aperture and resolution requirements and, therefore, we can use Equation (4-4) to calculate the combtooth length

$$L_c = \frac{t_c + g_0}{\sin(\theta_0)} \tag{4-6}$$

At the maximum deflection angle, the restoring torque from the torsional support equals the actuator torque, so

$$k_\theta = \frac{T_{max}}{\theta_0} = \frac{V_{max}^2 N\varepsilon_0 \left[\dfrac{(t_c - g_0)^2}{\sin^2 \theta_0} - d^2 \right]}{2g\theta_0} \qquad (4\text{-}7)$$

The resonant frequency of a mirror designed for steady-state beam steering is an important design variable that is related to the time required to steer the mirror from one position to another. Combining Equations (2-5), (2-21), (2-59) and (4-7) we calculate the resonant frequency ω_r

$$\omega_r = \sqrt{\frac{3V_{max}^2 N\varepsilon_0 \left[\dfrac{(t_c - g_0)^2}{\sin^2 \theta_0} - d^2 \right]}{8g\rho\, t_m W_m \left(\lambda N_{pixels} \right)^3}} \qquad (4\text{-}8)$$

Equation (4-8) shows that the maximum resonant frequency for a given scan angle, driving voltage, and optical resolution is increased if the combtooth thickness t_c increases. The thickness of the combteeth is effectively limited in two ways. First, for a given scan angle an increase in the combtooth thickness must be accompanied by an increase in the combtooth length (as shown in Equation (4-6)), but very long combteeth become laterally unstable at a critical applied voltage. The lateral stiffness of the combteeth is proportional to $\left(\dfrac{W_c}{L_c} \right)^3$, making it desirable to limit the length of the combteeth to avoid potential instability due to lateral combtooth bending. The critical voltage at which the combteeth bend laterally is also related to the misalignment between the fixed and moving combteeth.

Second, it is difficult to make very thick structures with low resonant frequencies in the STEC fabrication process. For thick structures, the aspect ratio of the torsion hinge must be high since the torsion-hinge thickness is the same as the combtooth thickness. Increasing the combtooth thickness while retaining the same resonant frequency requires progressively thinner torsion hinges, which can reduce the yield of the fabrication process.

From Equation (4-8), another way to increase the resonant frequency ω_r is to decrease the mirror thickness t_m (or the mirror effective density $\dfrac{\rho}{E}$) without changing the combtooth thickness t_c; this approach has been demonstrated in work by Nee and the author, which is discussed in section 2 and [2].

1.3.3.2   Resonant operation

The design limitations for high-speed resonant micromirror scanning are significantly different from those for steady-state beam steering. Whereas the moment-of-inertia of the mirror is an important design consideration for the resonant frequency (and therefore the response time) of steady-state beam-steering mirrors, the resonant frequency of mirrors targeted for resonant-scanning applications ultimately depends not on the moment-of-inertia, but on the damping, as shown in Chapter 2 section 2.2.

The scan angle of the STEC micromirror at resonance can be calculated by equating the energy input to the system per cycle and the energy lost per cycle. The total amount of energy added to the system through the actuator during each cycle E_{in} is

$$E_{in} = \oint T \dot{\theta} dt \tag{4-9}$$

where T is the torque and $\dot{\theta}$ is the angular velocity. The energy lost during each cycle E_{out} is

$$E_{out} = \oint b \dot{\theta}^2 dt \tag{4-10}$$

where b is the damping coefficient (the damping torque is $-b\dot{\theta}$). If the scan amplitude is stable, the energy added per cycle is equal to the energy lost per cycle. The torque of the STEC actuator is a function of the angle of the mirror, so the scan angle cannot be solved with Equations (4-9) and (4-10) using linear system theory. However, for systems with low damping operating at their resonant frequency, the motion is approximately sinusoidal and the maximum scan angle can be approximated as

$$\theta_{max} = \frac{2 \oint T \sin(\omega_r t) dt}{b \omega_r} \tag{4-11}$$

The actuator of the STEC micromirror only adds energy to the mirror during ½ of the scan because the moving and fixed combteeth are only capacitively coupled during the positive portion of the scan.

The form of Equation (4-11) shows that the torque, the damping coefficient, and the resonant frequency determine the amplitude of the motion θ_{max}. The maximum angle does not depend on the mirror moment-of-inertia (unlike the case described above for steady-state beam steering in Equation (4-8)) except insofar as the mirror geometry affects the damping coefficient b. Large reductions in damping – such as those made possible by operating in a vacuum – can reduce the actuator torque requirement for a given desired resonant frequency and scan angle. The damping coefficient depends strongly on mirror width and length, but is nearly independent of mirror thickness so that increasing the thickness of the micromirror does not increase the torque necessary for a given scan angle. It is necessary to have a stiffer spring in order to have the same resonant frequency for a thicker mirror, but when operated at resonance the maximum scan angle remains constant. Increasing the size of the mirror does increase the damping coefficient, and therefore decreases the achievable scan angle for a given torque and resonant frequency.

Equation (2-52) shows that the dynamic deformation is inversely proportional to the square of the mirror thickness; therefore, a thicker mirror will have less dynamic deformation. Since the scan angle at resonance remains constant with mirror thickness while the dynamic deformation decreases, thicker mirrors are capable of higher optical-resolution resonant scanning. Methods of decreasing the mirror moment-of-inertia, such as honeycomb support structures, do not improve the performance of mirrors in resonance because the effective density and Young's modulus both increase proportionally and thereby do not decrease the dynamic deformation (as shown in Equation (2-52)).

1.4 Results

Various STEC mirror geometries have been fabricated with sizes ranging from 200 µm up to 3 mm half-length, and resonant frequencies from 1 up to 61 kHz. SEMs of three STEC mirrors are shown in Figure 4-5 and Figure 4-6. Our measurements show that the STEC mirror excels in the critical performance criteria: resolution, scan speed, scan repeatability, size, power consumption, and reliability. This section discusses measurements of three of these performance criteria for one STEC mirror design.

1.4.1 Scan speed

STEC micromirrors with 550 µm diameter have been built with a wide variety of scan speeds, from 1 kHz up to more than 61 kHz – almost an order of magnitude higher than the resonant frequencies of commercially available optical scanners. Larger STEC mirrors have also been fabricated (up to 3 mm half length) with lower resonant frequencies. Figure 4-10 shows the measured frequency response of a 550 µm-diameter micromirror with 34 kHz resonant frequency.

1.4.2 Optical resolution

The surface deformation of the STEC micromirror was characterized using the stroboscopic interferometer reported in [27]. The largest dynamic deformation occurs at the end of the optical scan, where the angular acceleration, and therefore the inertial forces on the mirror, are largest. A surface-height map showing the total static nonplanarity of the mirror at this point is shown in Figure 4-11. The total deformation is less than 20 nm (considerably below the 82 nm Rayleigh limit for 655 nm light) and does not significantly reduce the optical resolution. The mirrors are expected to remain flat over a wide range of temperatures, as the thick single-crystal silicon mirrors are rigid enough to resist the stress caused by the thin reflective coating.

Figure 4-12 shows a photograph of the scan of a STEC mirror with a 655 nm laser pulsed synchronously with the mirror scan to generate 43 spots. The mirror scan is sinusoidal, so the timing of the laser pulses was adjusted across the scan to produce equally spaced spots. Figure 4-12 also shows a close-up image of three individual spots captured with a CCD camera. By fitting a Gaussian curve to each of these three spots, we were able to determine the full-width-half-max beam width and the distance between the spots. The spot size and separation at eight different regions across the scan give the measured total optical resolution of 350 pixels. The resolution of this 550 μm-diameter mirror with 24.9° optical scan and 655 nm laser light is demonstrated to be near the diffraction-limited resolution of 355 pixels from Equation (2-59). Therefore the STEC micromirror is capable of high-speed scanning with nearly ideal optical performance.

1.4.3   Power consumption

High-speed scanners require more torque than low-speed scanners to reach the same scan angle. In order to generate the torque necessary for large angle, high-frequency operation of the STEC micromirror, we use fairly high voltages. The 550 μm-diameter mirror with resonant frequency 34 kHz requires a 171 Vrms input sine wave for a total optical scan of 24.9°. To simplify mirror testing and operation, we use a small (1 cm^3) 25:1 transformer, allowing us to use a conventional 0-10 V function generator to drive the scanning mirrors with a sinusoidal waveform of amplitude up to 250 V. The use of the transformer also provides efficient power conversion, so the power consumption of the entire system can be much lower than that in systems requiring high-voltage power supplies and op-amps.

This power consumption is the sum of the dissipation in the drive electronics and the power dissipated by air and material damping. The power consumption due to damping is

$$P = \frac{1}{2}b\theta_o^2\omega_r^2 = \frac{1}{2}\frac{k_\theta}{Q}\theta_0^2\omega_r \qquad (4\text{-}12)$$

where k_θ is the torsional spring stiffness, b is the torque damping factor, θ_0 is the mechanical scan half angle (the total optical scan is $\pm 2\theta_0$), ω_r is the resonant frequency, and Q is the resonant quality factor. For the 34 kHz 550 µm-diameter mirror scanning 25° optical ($\pm 6.25°$ mechanical), the calculated stiffness $k_\theta = 3.94 \times 10^{-5}$ Nm/radiar, the measured resonant quality factor $Q = 273$, so the power consumption due to damping from Equation (4-12) is 0.18 mW. Vacuum packaging can be used to reduce the viscous damping, and thereby decrease the power consumption.

The measured power consumption is 6.8 mW, indicating that the majority of the power consumed is in charging and discharging the parasitic capacitances and in losses in the transformer.

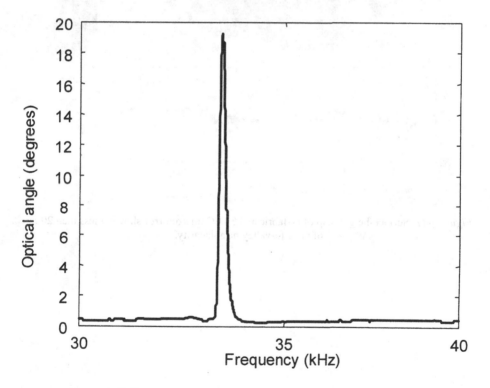

Figure 4-10. Frequency response of a STEC micromirror operated in air.

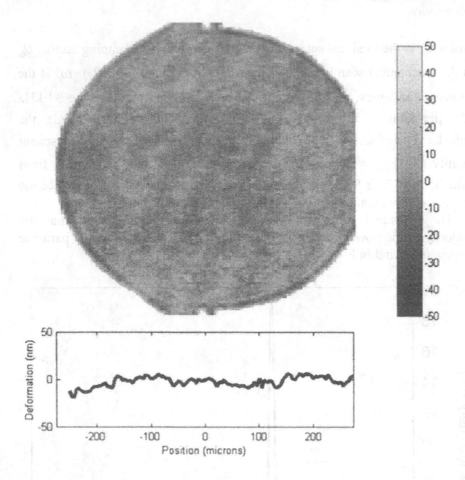

Figure 4-11. Surface-height map of as-fabricated STEC micromirror showing less than 20 nm of peak-to-valley nonplanarity.

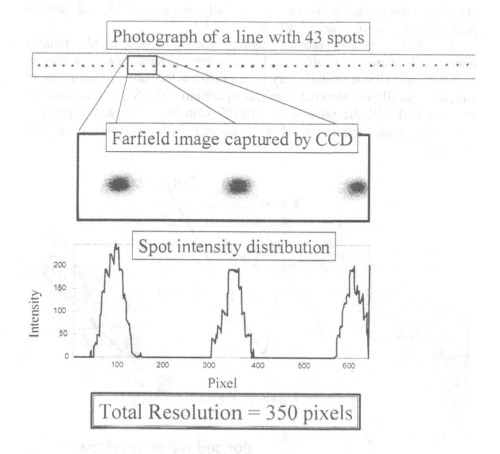

Figure 4-12. Optical resolution measurements from STEC micromirror showing nearly diffraction-limited optical performance.

1.5   STEC micromirror conclusions

The STEC micromirror enables significant performance enhancements over the surface-micromachined mirrors described in Chapter 2. The mirrors themselves are much thicker than surface-micromachined mirrors (20-100 μm for STEC mirrors versus 1-3 μm for surface-micromachined mirrors), thereby enabling high-speed resonant scanning with diffraction-limited optical performance over a wide range of temperatures (the bimorph effect does not produce significant mirror curvature for these thick mirrors [19]). These mirrors are easier to manufacture than surface-micromachined mirrors, eliminating the need for sequential polysilicon and silicon oxide deposition and etching by using a simple three-mask process. The mirror structure also allows the possibility of assembling 2-dimensional mirror scanning subsystems, as shown schematically in Figure 4-13, and vacuum packaging, as shown schematically in Figure 4-14.

The STEC mirrors provide highly repeatable and highly reliable operation because the single-crystal silicon torsion hinges have no play or backlash, and silicon exhibits very little plastic deformation in the range of stresses typically encountered in mirror operation. The STEC micromirrors promise to fulfill the promise of MEMS scanning mirrors: inexpensive, batch-fabricated, robust, reliable, high-speed, high-resolution optical scanners.

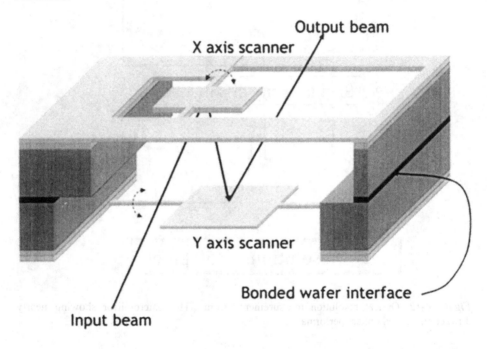

Figure 4-13. Schematic of 2-dimensional scanner subsystem using STEC micromirrors.

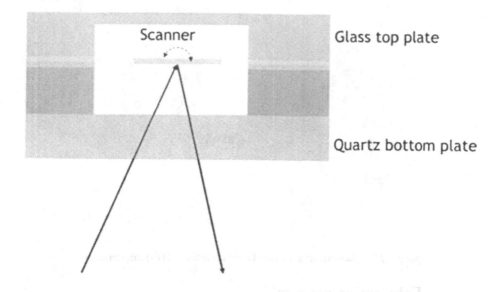

Figure 4-14. Schematic of vacuum package for STEC scanner.

2.   TENSILE OPTICAL SURFACE MICROMIRRORS

High-inertia mirrors – like the STEC mirrors described above – are well-suited for resonant-scanning applications, but require prohibitively large torque to achieve high-angle, high-speed, steady-state beam positioning. Low-inertia mirrors are capable of higher-speed steady-state beam steering. Dynamic deformation is not a leading concern for steady-state beam steering, but static mirror deformation is. The ideal mirror for steady-state beam steering is lightweight (thereby allowing high-speed beam steering), and stiff (to minimize static and dynamic deformation). The Tensile Optical Surface (TOS) micromirror was designed to meet both of these design goals.

The TOS mirror is shaped like a drum: a thin membrane is stretched across a stiff frame, as shown schematically in Figure 4-15. This mirror design effectively retains large combdrive thickness t_c but reduces the effective mirror thickness t_m, thereby increasing the achievable resonant frequency ω_r of Equation (4-8). The actuator of the TOS mirror is the same as that described for the STEC mirror in section 1. The moment-of-inertia of the TOS mirror, however, can be less than 10% that of the equivalent slab STEC mirror, resulting in a 3x increase in the resonant frequency.

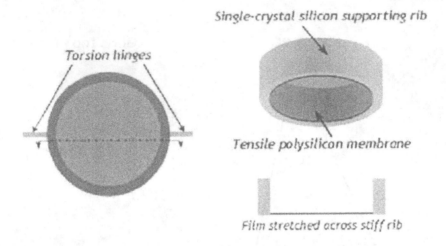

Figure 4-15. Schematic of a Tensile Optical Surface (TOS) micromirror.

2.1   Fabrication process

The fabrication process for the TOS micromirror is an adaptation of the STEC fabrication process flow, and the steps are listed in Appendix B. Figure 4-16 shows the STEC/TOS fabrication process flow. In step (c) a hole is etched into the top silicon layer, and in (d) a polysilicon membrane is deposited and etched, forming the TOS mirror structure. The other fabrication process steps are identical to the STEC process flow described in Appendix A and shown schematically in Figure 4-2. Careful attention must be paid to the membrane film stress, as too much tensile stress will cause the silicon rib to pull inward and will result in a concave optical surface. Compressive stress in the film will result in bowing of the membrane, and therefore a convex optical surface. A complete description of the STEC/TOS fabrication process flow, along with a detailed description of the fabrication issues, is given by Jocelyn Nee [2, 11].

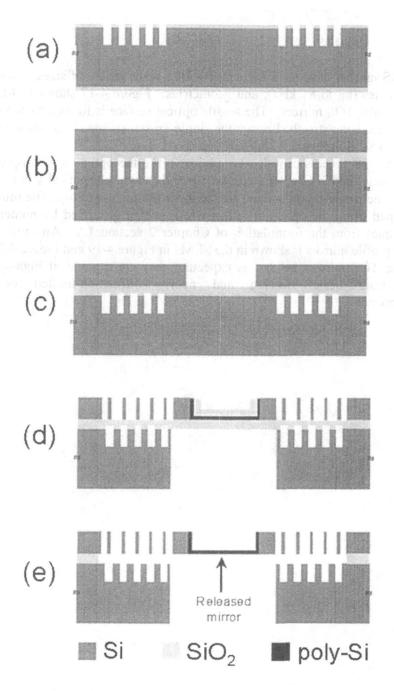

Figure 4-16. Fabrication process flow for STEC/TOS mirror.

2.2   Results

TOS mirrors have been fabricated with a wide variety of sizes, resonant frequencies (up to 81 kHz), and geometries. Figure 4-17 shows SEMs of two circular TOS mirrors. The tensile optical surface is formed by the thin polysilicon layer stretched across the single-crystal silicon rib, as shown in a close-up SEM in Figure 4-18.

The TOS process can also be used to make the optimum-width-profile mirror described in Chapter 2 section 1.6 and shown schematically in Figure 2-16. The single-crystal silicon rib for these optimum-width-profile mirrors is shaped using the optimum mirror-width profile generated by numerical techniques from the formulation of Chapter 2 section 1.6. An optimum-width-profile mirrors is shown in the SEMs in Figure 4-19 and Figure 4-20.

The TOS mirrors behave as expected; they are capable of high-speed steady-state beam steering and nearly diffraction-limited optical performance.

Figure 4-18. Close-up SEM of mirror rib and membrane of TOS mirror.

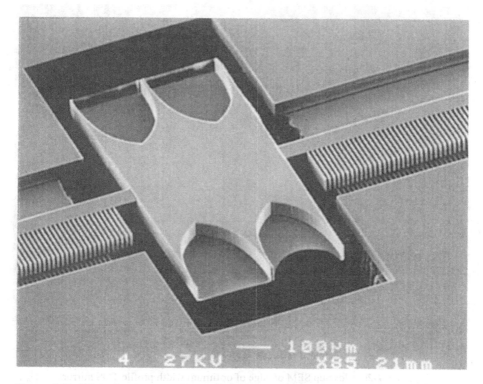

Figure 4-19. SEM of optimum width profile mirror with TOS rectangular reflective surface.

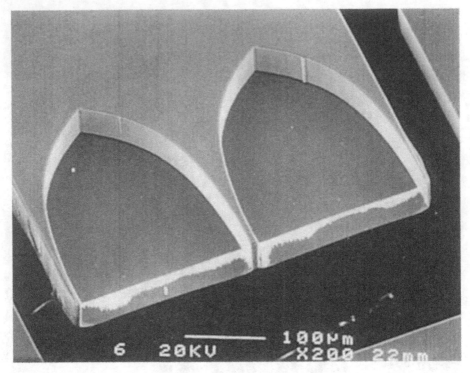

Figure 4-20. Close-up SEM of edge of optimum width profile TOS mirror.

Chapter 5

Example Application of Scanning Mirrors
Raster-Scanning Projection Video Display

1.   INTRODUCTION

Micromachining technology can enable a new generation of portable projection-video displays that are lower in weight, power consumption, and cost than displays that are available today.   Emerging micromachined projection video display technologies fall into three general categories:

1. Two-*dimensional arrays*. Texas Instruments' Digital Light Processor (DLP) display [93] and reflective LCD displays use one reflective element for each pixel. The reflected elements are imaged onto the projection screen to form the two-dimensional image.

2. *Scanned one-dimensional arrays*. Silicon Light Machines' grating-light-valve display uses one scanning mirror and a one-dimensional array of light modulators [94] to generate a two-dimensional image [95]. Another scanned one-dimensional array display has been made using an array of switched laser light sources and one scanning mirror [96].

3. *Raster-scanned*. Raster-scanned displays use two mirrors scanning in orthogonal directions or a single gimbaled mirror scanning in two dimensions, and a modulated light source to generate a 2-D image [97]. Such displays have been demonstrated for direct virtual projection displays [98] and laser projection displays [99].

Each of the three display architectures enumerated above has its advantages. A 2-D array of mirrors, such as the Texas Instruments DLP, is less sensitive to the mirror surface curvature and scan irregularities that are detrimental to scanned 1-D array and raster-scanned displays. The 2-D pixel-array displays, however, require successful fabrication of *mxn* elements for a display with m rows and n columns. The scanned 1-D array displays, on the other hand, require m elements, and the raster-scanned displays requires only one or two elements. Raster-scanned displays are also potentially smaller and less costly than 2-D and scanned 1-D displays, making them attractive for portable display applications.

In this chapter we describe a raster-scanned video display system, similar to the display architecture used in modern cathode-ray tube (CRT) displays, which are ubiquitous in television and desktop-computing applications. CRT displays make use of a modulated electron beam that is deflected by magnetic field. The electrons impact the phosphor-coated front of the CRT and excite the phosphors, causing them to emit light. The micromachined raster-scanning video display, on the other hand, scans the modulated light beam directly as opposed to scanning the electron beam and relying on the excitation of a phosphor to generate light.

A bulk-micromachined raster-scanning display with piezoelectric actuators has been previously reported in the literature [100]. Surface-micromachined mirrors with monolithically integrated electrostatic-combdrive actuators have been demonstrated to display single raster-scanned video frames [5], [101]. The work presented here extends the results from [97] and [101] to include full-motion video images.

2. DESIGN AND FABRICATION

The MEMS video display we constructed uses two electrostatic-combdrive-actuated micromirrors fabricated at the publicly available MCNC MUMPS foundry process [33]. Scanning-electron micrographs (SEMs) of the two micromirrors are shown in Figure 5-1. The MUMPS process includes two structural layers of polycrystalline silicon: the 2 μm-thick POLY1 layer, and the 1.5 μm-thick POLY2 layer. The electrostatic combdrives are fabricated with 3.5 μm-thick (POLY1+POLY2) polysilicon. The mirrors and torsion hinges are fabricated in the POLY2 layer. The mirrors are attached with torsion hinges to the support frame, which are connected to the substrate with polysilicon pin hinges. The micromirror assemblage is folded up from the surface after fabrication to hold the mirror perpendicular to the substrate plane. More detail on the design and fabrication of fold-up surface micromachined scanning mirrors has already been published [59], and is discussed in Chapter 2.

Figure 5-1. SEM of the 550 μm diameter line-scan mirror, and the 500 x 764 μm rectangular frame-scan mirror used for the video display.

The two mirrors used in the video display must satisfy different requirements. The frame-scan mirror should perform a linear scan at a low frequency (60 Hz for a typical display refresh rate), whereas the line-scan mirror should scan at a high frequency (greater than 28 kHz for a 640x480 VGA display). Both mirrors must have very stable, repeatable scans (better than a fraction of a pixel) for good image stability and high resolution.

The frame-scan mirror was designed to provide a large scan angle when operated below its mechanical resonant frequency. The distance between the combdrive linkage and the torsion hinge on the frame-scan mirror is small in order to create a large mirror scan with a small combdrive displacement.

The line-scan mirror is driven at resonance to take advantage of the resonant scan amplification. The line-scan mirror torsional hinge is in the middle of the mirror to reduce the rotational moment-of-inertia of the mirror and to maximize its mechanical resonant frequency.

3.   VIDEO SYSTEM

The standard video card output consists of five signals plus ground: Vsync digital pulse at the start of each video frame, Hsync digital pulse at the start of each line, and three 1.5 volt peak-to-peak video signals (one each for red, green, and blue). The interface between the output of the computer video card and the MEMS video display uses the Vsync to drive the frame-scan mirror, Hsync to drive the line-scan mirror, and the red video signal to modulate the laser diode. For our demonstration system, we adjusted the video frame rate (also called the refresh rate) to 20 Hz so that the line-scan frequency matched the line-scan mirror resonant frequency (6.2 kHz).

Figure 5-2 shows the video-signal timing, and the waveforms used to drive the scanning mirrors. The force exerted by the electrostatic-combdrive actuators is proportional to the square of the input voltage, so the drive signal for the frame-scan mirror is the square root of a sawtooth wave (generating a linear scan). The line-scan mirror is driven to excite the resonant frequency (the drive voltage is a sine wave at half the resonant frequency). The laser diode modulation signal is active during the left-to-right scan of the line-scan mirror.

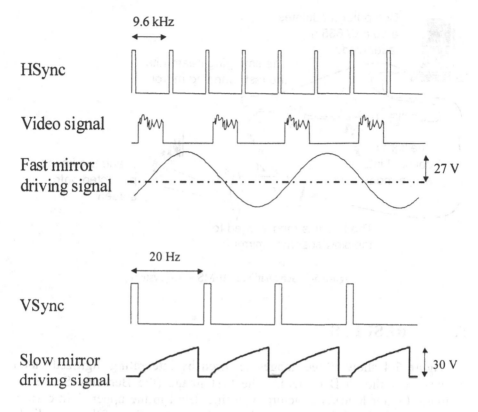

Figure 5-2. Video signal timing.

4. OPTICAL SYSTEM

The optical system is shown schematically in Figure 5-3. The light source is a 10 mW 635 nm laser diode, which is driven with a voltage-controlled current source connected to the red video signal. To give the scanned beam a Gaussian profile, the light from the laser diode is coupled into a short length of single-mode 635-nm optical fiber (the coupled power is 3 mW). The Gaussian profile results in close to optimum system resolution. The light emitted from the optical fiber is directed through two lenses, forming a waist at the line-scan mirror with a $1/e^2$ radius of 250 microns. The line-scan mirror is then imaged onto the frame-scan mirror using two lenses in a 4f configuration, as shown in Figure 5-3. The light reflected from the frame-scan mirror is focused onto a CCD camera using an additional lens. The surface micromachined mirrors have significant static curvature after release, which is compensated by adjusting the position of the output lens to obtain optimal focus on the CCD image plane.

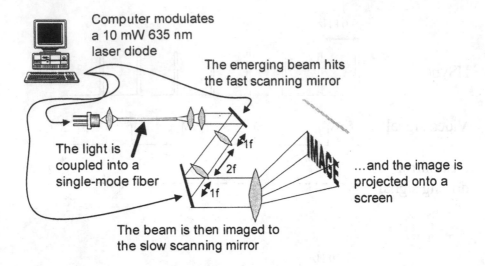

Figure 5-3. Schematic of MEMS video system.

5.   RESULTS

Figure 5-4 shows three images acquired by integrating digitized video images from the CCD camera. The first image (the Berkeley Sensor & Actuator Center logo) is a picture with fine detail in the upper right corner. The image of a hand demonstrates that analog modulation of the laser diode provides grayscale capability.

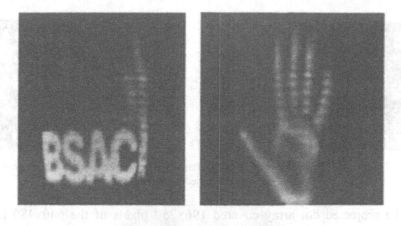

Figure 5-4. Recorded images from the video display.

By displaying a 5x8 array of dots on the MEMS projection display (shown in Figure 5-5), we were able to quantify the performance of the display. As is evident in Figure 5-5, the array of spots is not square, indicating some curvature of field produced by the optical system. The spot size also varies across the image, indicating variations in optical resolution across the field of view.

Figure 5-5. Image of a 5x8 square array of spots used to measure resolution – as programmed (left) and as projected (right).

The projected dot array covered 196x250 pixels of the 640x480 pixel CCD camera, and the resolution was calculated by measuring the size of each dot on the CCD camera. At the highest-resolution portion of the projected image, the full-width-half-max (FWHM) spot (the preferred method of determining CRT resolution [102]) covered 2.7 x 2.7 pixels on the CCD camera. At the top of the image, the FWHM spot was 7x7 CCD pixels. The average spot size across the entire MEMS display was 4.8 CCD pixels, resulting in a total number of resolvable spots across the projected image of 41 x 52 pixels.

The maximum achievable resolution is limited by diffraction from the 550 µm diameter mirror aperture. For a 15° optical mirror scan, the diffraction-limited resolution is on the order of 180x180 pixels. Clearly the system is not diffraction limited, but rather limited by the static and dynamic mirror curvature. The defocus term of the static deformation is corrected by adjusting the position of the projection lens, but the higher-order aberrations (including cylindrical curvature of the mirror surface) were not corrected in the display system presented here.

Dynamic mirror deformation causes a significant reduction in the optical resolution. Even with optimal correction for the static curvature, the far-field spot at the edge of the line scan is more than 10 times larger than the diffraction-limited spot size [27]. As a result, image resolution is especially poor in the vicinity of the edge of the display. Stiffer mirrors are necessary to reduce the dynamic deformation.

Scan irregularity and electrical noise, including timing differences between the laser-modulation signal and the mirror driving signal, could reduce image stability and resolution. The captured video images contain information about the position of each spot of the 5x8 dot array at different times, and this information was used to determine the irregularity of the mirror scan. A Gaussian peak was fitted to each dot in both the vertical and horizontal directions; the variation between the best-fit Gaussian positions in sequential frames indicates the irregularity of the mirror scan. Analysis of three seconds (60 frames) of video data indicates mirror-scan irregularity to be less than 0.25 pixels on the CCD camera. A realistic estimate of the uncertainty of this measurement is ±0.25 pixels, which sets an upper bound of the scan irregularity of 0.5 CCD pixels, corresponding to a resolution of 393x499 pixels – well above the diffraction limit. From these considerations, we conclude that the mirror-scan irregularity and electrical noise will not be the resolution-limiting factor for this system.

Chapter 6

Conclusions

Micromachined scanning mirrors are interesting for a wide variety of applications because of their potential low cost, high speed, low power consumption, and reliability. These mirrors can offer significant advantages over macro-scale mirrors, but the fundamental limitations of scanning mirrors have not been widely discussed. In this dissertation we have derived the range of performance, in terms of resonant frequency and optical resolution, that can be achieved using scanning mirrors, and have shown that MEMS-scale scanning mirrors offer fundamental advantages over the macro-scale scanning mirrors commercially available today.

1. SCALING LAWS

The scaling laws for resonant scanning mirrors can be derived from the theory we have given after making certain assumptions. For the scaling laws presented here, the following constraints have been applied:
1. The scan angle is fixed. This implies that the torque of the system is sufficient to achieve the desired scan angle at the desired resonant frequency.
2. The mirror thickness and length scale proportionally.
3. The wavelength of light is fixed.

With these constraints, Equations (2-59) and (2-64) show that the scaling laws for resonant scanning mirrors are:

$$f \propto \frac{1}{L^{3/2}}$$
$$N_{pix} \propto L$$

(6-1)

Equation (6-1 shows that as the mirror is scaled to smaller dimensions, the optical resolution of the system is reduced, but the frequency is increased, implying that MEMS-scale resonant scanning mirrors will be particularly useful for high-speed scanning in applications that require moderate resolution. Since many optical systems (including laser printers and video displays) can be built with multiple scanners working in parallel to increase resolution, MEMS-scale optical scanners will provide higher-speed performance than is achievable using macro-scale resonant scanners.

2.   FABRICATION TECHNOLOGIES

The fabrication processes used to make MEMS devices offer some significant advantages over conventional fabrication processes.   MEMS devices can:
– Have features ranging from nanometers or micrometers up to
   millimeters, and these features can be very precisely controlled;
– Be batch fabricated, allowing hundreds or thousands of devices to be
   made at the same time, thereby dramatically reducing the cost of
   manufacture; and
– Be made with a variety of different materials, some of which offer
   excellent mechanical properties without the voids and inclusions often
   found in macro-scale mechanical structures.
However, the fabrication processes often limit the achievable device geometries. The surface-micromachining fabrication processes described in Chapter 3 are quite useful for making a wide variety of complex systems, but the limited thickness of the thin-film LPCVD polysilicon films makes this fabrication process unsuitable for high-speed, high-resolution scanning micromirrors. The thin-film mirrors suffer significant degradation in optical resolution due to dynamic deformation even at relatively modest resonant frequencies.

In addition, scanning mirrors made using the surface-micromachining fabrication processes described in Chapter 3 must be folded up from the substrate after they are released from their supporting oxide layers. Yield on this process can be low, and the hinges employed for these fold-up structures can lead to hinge "play" and backlash that are degrading to scanning mirror performance.

Even with these limitations, surface-micromachined mirrors have a number of potential applications, as demonstrated by the raster-scanning video display presented in Chapter 5. The ready availability of surface-micromachining foundry services makes these mirrors attractive for proof-of-concept demonstrations, even though custom fabrication processes will often provide higher-performance devices at lower cost with moderate to high production volumes.

The theoretical framework for understanding the mechanical dynamics of scanning mirrors, as presented in Chapter 2, provides a set of guidelines for the design of custom fabrication processes for scanning micromirrors. The micromirrors should be thick, flat structures without the need for post-fabrication assembly. The mirrors should be directly coupled with an integrated high-torque actuator that is attached to the mirror with a rigid mechanical connection to apply a torque directly to the mirror. Using these guidelines, we have designed the Staggered Torsional Electrostatic Combdrive (STEC) micromirrors and the Tensile Optical Surface (TOS) micromirrors presented in Chapter 4 for high-speed, high-resolution scanning. These mirrors demonstrated the promise of MEMS scanning mirrors: they are low-cost, high-speed, high-resolution, reliable, and consume little power.

We have demonstrated mirrors capable of diffraction-limited optical scanning at speeds greater than 30 kHz – significantly faster than that demonstrated with conventionally fabricated scanning mirrors. The STEC micromirrors are fabricated in a relatively simple three-mask process, have an integrated high-torque actuator that can provide capacitive angle sensing capability, and require no post-fabrication assembly.

3. FUTURE WORK

Micromirrors for high-speed resonant scanning and steady-state beam steering can be fabricated for commercial applications with existing technology, but further research and development will clearly enhance their performance. Directions of future work can include:

– **Mirror packaging**. Mirror packaging is a critical issue for the successful realization of low-cost MEMS scanning mirrors. The development of low-cost, hermetically sealed packaging for MEMS mirrors will require consideration of the reliability of the MEMS mirrors, including development of an understanding the failure modes of the mirrors in humid environments. Thermal expansion coefficient differences between the MEMS mirror and the package will cause stress on the mirror chip that may affect the both the sort-term performance and the long-term reliability of the mirror, so developing a thermally matched package will be critical.

- **Optical coatings**. The mirrors presented in this work used thin evaporated layers of chrome and gold to achieve greater than 90% mirror reflectivity for visible and infrared light. More complex multi-layer optical coatings could be used to achieve higher reflectivity and wavelength selectivity, which could be useful for some applications. The successful implementation of these multilayer mirror coatings will depend on the interaction of the coating material with the mechanical properties of the torsion hinge, and the robustness of the coating to the liquid hydrofluoric acid release process.

- **Position sensing**. Mirror position sensing for closed-loop mirror position control will be useful for both resonant scanning and steady-state beam-steering applications. The electrostatic actuators used for both the surface-micromachined mirrors and the STEC micromirrors can be used for capacitive position sensing. Future work could evaluate the effectiveness of this position sensing method where the combdrive is used for capacitive sensing, or an additional interdigitated comb structure is used exclusively for capacitive sensing.

- **Optimal width-profile TOS mirrors**. The TOS micromirrors can be used to make mirrors with the optimal width profile, as described in Chapter 2 section 1.7. Mirrors with this shape have been fabricated and tested, but the dynamic deformation equations derived herein have not been compared to measured dynamic deformation. Future work could validate this dynamic deformation model and show the limits of steady-state beam steering using these optimally shaped mirrors.

- **Variable thickness mirrors**. The formulation for mirror dynamic deformation assumed a constant mirror thickness. Significant reductions in dynamic deformation would be possible using varying mirror thickness, but these mirrors cannot be fabricated using the planar MEMS fabrication processes described here. The mathematical formulation for mirror dynamic deformation could be extended to mirrors of varying thickness, and optimal mirror shapes can be derived using these assumptions. Innovative fabrication processes can then be used to fabricate mirrors with reduced dynamic deformation.

Appendix A

STEC FABRICATION PROCESS FLOW

0.0 Starting Wafers: <100>, 500-550 μm thickness, n-type prime.

1.0 Thermal Oxidation

    1.1 Standard clean (sink8/sink6).

    1.2 Wet oxidation (Tylan3/4)

        (a) Handle (bottom) wafers: SWETOXB, 1000°C, 25 min. Target: 0.2 μm.

        (b) Device (top) wafers: SWETOXB, 1100°C, 1.1 hrs. Target: 1.5 μm.

2.0 BURIED Feature Definition

    2.0 Starting Wafers: Handle (bottom) wafers with 0.2 μm silicon dioxide.

    2.1 Lithography with BURIED mask: Standard Litho Module #1.2.

2.2 Silicon dioxide etch (lam2), Standard Etch Module #2: Etch time = 30 sec.

2.3 Deep reactive ion etch (DRIE) of silicon (STS), Standard Etch Module #4: Etch time = 45 min. Target depth = 100 μm.

2.4 Label wafer on backside

2.5 Strip photoresist in 90°C PRS3000 (sink5), rinse and dry (spindryer3) using program 2.

3.0 SOI Wafer Formation

3.1 Standard clean (sink8/sink6) of all wafers, without HF dip, and with long (>30 min) piranha dip to remove sidewall polymer resulting from DRIE process.

3.2 Bond wafers:

(a) Precautions: Wear poly-d gloves on both hands. Perform bonding under HEPA filters (at sink6 or furnace area).

(b) Using vacuum wand, pick up patterned handle (bottom) wafer and place in palm, polished side up.

(c) Using vacuum wand, pick up device (top) wafer and place in palm, polished side facing handle wafer. Do not let wafers touch each other.

(d) Align major flats of the two wafers.

(e) Position wafers ~5 mm apart.

(f) Press in center of wafers to bond center of wafers, then let the wafers come together.

3.3 Standard clean (sink6 only) of bonded wafers without HF dip.

3.4 Anneal (Tylan3/4) for bond integrity: SWETOXB, 1100°C, 1 hr.

3.5 Inspect bond with IR light/camera.

3.6 Grind and polish wafer. Top silicon thickness = 50 μm.

3.7 Standard clean (sink8/sink6).

3.8 Wet oxidation (Tylan3/4): SWETOXB, 1100°C, 25 min. Target: 1 μm.

4.0 Open Alignment Dice

4.1 Lithography with blank mask: Standard Litho Module #1.2, expose only the two alignment dice.

4.2 Silicon dioxide etch (lam2), Standard Etch Module #2: Etch time = 3 min.

4.3 Deep reactive ion etch (STS), Standard Etch Module #4: Etch time = 20 min.

4.4 Strip photoresist in 90°C PRS3000 (sink5), rinse and dry (spindryer3) using program 2.

5.0 FRONTSIDE Feature Definition

5.1 Lithography with FRONTSIDE mask: Standard Litho Module #1.2. Align to BURIED pattern.

5.2 Silicon dioxide etch (lam2), Standard Etch Module #2: Etch time = 3 min.

5.3 Strip photoresist in 90°C PRS3000 (sink5), rinse and dry (spindryer3) using program 2.

6.0 HOLE Feature Definition on backside of wafer

6.1 Lithography with HOLE mask: Standard Litho Module #2.

6.2 Silicon dioxide etch (lam2), Standard Etch Module #2: Etch time = 3 min.

6.3 Bond to dummy wafer for through wafer STS etch

   (a) Apply photoresist (svgcoat1/svgcoat2) to front side of wafer, single coat of g-line (Program 4), no bake, target: ~2 μm of resist.

   (b) Apply photoresist (svgcoat1) to dummy wafer, single coat of thick gline (Program 8), no bake, target: ~9 μm of resist.

   (c) Bond wafer to dummy, photoresist-coated sides facing each other.

   (d) Hard bake, 120°C oven in vacuum (primeoven), ~30 min.

6.4 Deep reactive ion etch (DRIE) of silicon (STS), Standard Etch Module #4: Etch time = 3.25 hrs.

6.5 Strip photoresist in 90°C PRS3000 (sink5), separate wafer from dummy, rinse and dry (spindryer3) using program 2.

7.0 FRONTSIDE Etch

7.1 Bond to dummy wafer for STS etch

(a) Apply photoresist (svgcoat1) to dummy wafer, single coat of thick gline (Program 8), no bake, target: ~9 μm of resist.

(b) Bond front side of wafer to dummy.

(c) Hard bake, 120°C oven in vacuum (primeoven), ~30 min.

7.2 Deep reactive ion etch (DRIE) of silicon (STS), Standard Etch Module #4: Etch time = 18 min.

8.0 Post Process

8.1 Coat wafer with photoresist for protection of top surface.

8.2 Dice wafer (disco). Note: Do not dice into dummy wafer.

8.3 Strip photoresist in 90°C PRS3000, separate dice from dummy wafer, rinse, air dry.

8.4 Clean in SC-1 solution (RCA organic clean), 70°C, 20-30 min. Rinse, air dry.

8.5 Release structures

(a) Immerse in concentrated HF. Etch time: 2-3 min.

(b) Rinse with DI water until pH=7.

(c) Replace DI water with isopropyl alcohol.

(d) Dry in air.

8.6 Evaporate Au (nrc) onto mirror surface. Thickness: 500 Å.

Appendix B

TOS Fabrication Process Flow

(insert just before step 6.0 in STEC process)

1.0 MEMBRANE Feature Definition

    1.1 Lithography with MEMBRANE mask: Standard Litho Module #1.2. Align to FRONTSIDE pattern.

    1.2 Silicon dioxide etch (lam2), Standard Etch Module #2: Etch time = 3 min.

    1.3 Deep reactive ion etch (DRIE) of silicon (STS), Standard Etch Module #4: Etch time = 18 min.

    1.4 Ash (technics-c) to remove sidewall polymer resulting from DRIE process, O2, 300 W, 10 min.

    1.5 Strip remaining photoresist in 90°C PRS3000 (sink5), rinse and dry (spindryer3) using program 2.

2.0 Polysilicon and LTO Deposition

2.1 Standard clean (sink8/sink6). Does not need to be repeated in between furnace runs, as long as wafers are transported in a clean box.

2.2 Amorphous silicon deposition (Tylan16) 16FUPLYA, 590°C, 300 mtorr, SiH4=100 sccm, 2.5 hrs., target: 8000 Å.

2.3 Low temperature oxide (LTO) deposition (Tylan20) VDOLTOD, 450°C, 300 mtorr, SiH4=100 sccm, 1.5 hrs, target: 1 μm

2.4 Anneal (Tylan3/4), HIN2ANNL, Temp = 750°C-950°C depending on desired polysilicon stress, time=1 hr.

3.0 PROTECTION Oxide Feature Definition

3.1 Lithography with PROTECTION mask: Standard Litho Module #1.2. Align to FRONTSIDE pattern.

3.2 Silicon dioxide etch (lam2), Standard Etch Module #2: Etch time = 3 min.

3.3 Strip photoresist in 90°C PRS3000 (sink5), rinse and dry (spindryer3) using program 2.

4.0 Strip Backside Films

4.1 Silicon dioxide etch (lam2), Standard Etch Module #2: Etch time = 3 min.

4.2 Deep reactive ion etch (DRIE) of silicon (STS), Standard
Etch Module #4: Etch time = 1 min.

References

[1] K. E. Petersen, "Silicon torsional scanning mirror," *IBM Journal of Research and Development.*, vol. 24, no. 5, pp. 631–637, 1980.

[2] J. Nee, "Hybrid surface-/bulk-micromachining processes for scanning micro-optical components," Ph.D. dissertation, University of California, Berkeley, California, December 2001.

[3] M. R. Hart, R.A. Conant, K.Y. Lau, and R.S. Muller, "Stroboscopic interferometer system for dynamic MEMS characterization," UC Berkeley ERL Report, May 1 2000.

[4] M.A. Helmbrecht, "Micromirror Arrays for Adaptive Optics," Ph.D. dissertation, University of California, Berkeley, California, May 2002.

[5] C.W. Chang, "Magnetically Actuated Scanning Microplatforms for Intravascular Ultrasound Imaging," Ph.D. dissertation, University of California, Berkeley, California, May 2002.

[6] M. Daneman, "Micromachined photonic devices and systems on silicon," Ph.D. dissertation, University of California, Berkeley, California, November 1996.

[7] M.-H. Kiang, "Micro-optical devices for communications and beyond -- the days before and after silicon micromachining," Ph.D. dissertation, University of California, Berkeley, California, December 1997.

[8] R.A. Conant, J. Nee, K.-Y. Lau, and R.S Muller, "Dynamic deformation of scanning mirrors," in *2000 IEEE/LEOS International Conference on Optical MEMS (MOEMS 2000)*, 2000, Kauai, Hawaii, pp. 49-50

[9] R.A. Conant, J.T. Nee, M. Hart, O. Solgaard, K.-Y. Lau, and R.S. Muller, "Reliability and robustness of micromachined scanning mirrors," in *Proceedings of the 1999 IEEE/LEOS International Conference on Optical MEMS*, 1999, Mainz, Germany.

[10] R.A. Conant, J.T. Nee, K.-Y. Lau, and R.S. Muller, "A fast flat scanning micromirror," *IEEE Solid-State Sensor and Actuator Workshop Technical Digest*, June 2-4, 2000, Hilton Head Island, SC, pp. 6-9.

[11] J.T. Nee, R.A. Conant, K.-Y. Lau, and R.S. Muller, "Stretched-Film Micromirrors for Improved Optical Flatness," *Proceedings of the IEEE International Conference on Micro Electro Mechanical Systems (MEMS 2000)*, 2000, Miyazaki, Japan, pp.704-709

[12] R.A. Conant, P.M. Hagelin, U. Krishnamoorthy, M. Hart, O. Solgaard, K.Y. Lau, and R.S. Muller, "A raster-scanning full-motion video display using polysilicon micromachined mirrors," *Sensors & Actuators A (Physical)*, vol. 83, no. 1-3, pp 291-296, May 2000.

[13] R.A. Conant, P.M. Hagelin, U. Krishnamoorthy, O. Solgaard, K.-Y. Lau, and R.S. Muller, "A Full-Motion Video Display Using Micromachined Scanning Micromirrors," in *Proceedings of the International Conference on Solid-State Sensors and Actuators (Transducers '99)*, June 1999, Sendai, Japan, pp. 376-379.

[14] P.M. Hagelin, U. Krishnamoorthy, R.A. Conant, R.S. Muller, K.-Y. Lau, and O. Solgaard, "Integrated micromachined scanning display systems," in *Proceedings of the*

18th Congress of the International Commission for Optics, August 1999, San Francisco, CA, pp. 472-473

15] S.S. Rao, Mechanical Vibrations, 3rd edition, Reading, Mass: Addison-Wesley, 1993.

16] J.E. Mehner, J.D. Gabbay, and S.D. Senturia, "Computer-aided generation of nonlinear reduced-order dynamic macromodels – II: stress-stiffened case," *Journal of Microelectromechanical Systems*, vol. 9, no. 2, pp. 270-278, June 2000.

17] R.J. Roark, *Roark's Formulas for Stress and Strain*, Warren C. Young. 6th edition, New York: McGraw-Hill, 1989.

18] H. Miyajima, N. Asaoka, M. Arima, Y. Minamoto, K. Murakami, K. Tokuda, and K. Matsumoto, "A durable, shock-resistant electromagnetic optical scanner with polyimide-based hinges," *Journal of Microelectromechanical Systems*, vol. 10, no. 3, September 2001.

19] K. Cao, W. Liu, and J.T. Talghader, "Curvature compensation in micromirrors with high-reflectivity optical coatings," *Journal of Microelectromechanical Systems*, vol. 10, no. 3, pp. 400-417, Sept. 2001.

20] Y.-H. Min, and Y.-K. Kim, "Modeling, design, fabrication, and measurement of a single layer polysilicon micromirror with initial curvature compensation," *Sensors & Actuators A (Physical)*, vol. 78, pp. 8-17, 1999.

21] J. Comtois, "Surface-micromachined polysilicon MOEMS for adaptive optics," *Sensors & Actuators A (Physical)*, vol. 78, pp. 54-62, 1999.

22] D.L. Hetherington and J.J. Sniegowski, "Improved polysilicon surface-micromachined micromirror devices using chemical-mechanical polishing," *Proceedings of the SPIE*, vol. 3440, 1998, pp. 148-153.

23] M.A. Michalicek, J.H. Comtois, C.C. Barron, "Design and characterization of next-generation micromirrors fabricated in a four-level, planarized, surface-micromachined polycrystalline process," *Proceedings of Innovative Systems In Silicon*, 2nd Ed., IEEE Press, pp. 144-154, 1997.

[24] P.J. Brosens, "Dynamic Mirror Distortions in Optical Scanning," *Applied Optics*, vol. 11, no 12 Dec. 1972.

[25] J. DenHartog, *Mechanical Vibrations* (McGraw-Hill, New York, 1956)

[26] S. Timoshenko, *Theory of Elasticity*, New York: McGraw-Hill, 1970.

[27] M. Hart, R.A. Conant, K.-Y. Lau, and R.S. Muller, "Stroboscopic Interferometer System for Dynamic MEMS Characterization," *Journal of Microelectromechanical Systems*, vol. 9, pp. 409-418, December 2000

[28] Brantly, W., "Calculated Elastic Constants for Stress Problems Associated with Semiconductor Devices," *Journal of Applied Physics*, vol. 44, pp. 534-535, 1973.

[29] J. W. Goodman, *Introduction to Fourier Optics*, Second ed, New York: McGraw-Hill, 1996.

[30] L. Beiser, and R.B. Johnson, "Scanners," in *Handbook of Optics*, vol. 2, M. Bass, Ed. New York: Mc-Graw Hill, 1995.

[31] H.C. Nathanson, W.E. Newell, R.AA. Wickstrom, and J.R. Davis, "The resonant gate transistor", *IEEE Transactions on Electron Devices*, ED-14, 117-133, 1967.

[32] R.T. Howe, and R.S. Muller, "Polycrystalline silicon micromechanical beams", in *Proceedings of the Spring Meeting of the Electrochemical Society*, 1982, p. 184-185.

[33] D. Koester, R. Majedevan, A. Shishkoff and K. Markus, "Multi-user MEMS processes (MUMPs) introduction and design rules," Rev. 4, MCNC MEMS Technology Applications Center, Research Triangle Park, NC, USA, July 1996.

[34] K.S.J. Pister, "Hinged polysilicon structures with integrated CMOS TFT's," *Technical Digest of the 1992 Solid State Sensor and Actuator Workshop*, 1992, Hilton Head Island, SC, pp. 136-139.

[35] K.S.J. Pister, M.W. Judy, S.R. Burgett, and R.S. Fearing, "Microfabricated hinges," *Sensors and Actuators A*, vol. 33, pp. 249-256, 1992.

[36] L. Fan, and M.C. Wu, "Self-assembled micro-XYZ stages for optical scanning and alignment," in Proceedings of the 10th Annual Meeting of the IEEE Lasers and Electro-Optics Society, San Francisco, CA, 1997, pp. 266-267.

[37] L. Fan and M. C. Wu, "Two-dimensional optical scanner with large angular rotation realized by self-assembled micro-elevator," in *Digest of 1998 IEEE/LEOS Summer Topical Meeting*, Monterey, CA, July 1998, pp. II/107-108.

[38] P.M. Hagelin, and O. Solgaard, "Optical raster-scanning displays based on surface-micromachined polysilicon mirrors". *IEEE Journal of Selected Topics in Quantum Electronics*, vol.5, (no.1), Jan.-Feb. 1999. p.67-74.

[39] N.C. Tien, O. Solgaard, M.-H. Kiang, M. Daneman, K.-Y. Lau, and R.S. Muller, "Surface micromachined mirrors for laser-beam positioning," *Journal of Microelectromechanical Systems*, vol. 5, pp. 159-165, 1996.

[40] M.S. Rodgers, J.J. Sniegowski, J.J. Allen, S.L. Miller, J.H. Smith, P.J. McWhorter, "Intricate mechanisms-on-a-chip enabled by 5-level surface micromachining," ," in *Proceedings of the International Conference on Solid-State Sensors and Actuators (Transducers '99)*, June 1999, Sendai, Japan, pp.990-993.

[41] J. T. Butler, V. M. Bright, and J. R. Reid, "Scanning and rotating micromirrors using thermal actuators" in *Proceedings of the SPIE*, vol. 3131, 1997, pp. 134–144.

[42] M. Last, and K.S.J. Pister, "2-DOF actuated micromirror designed for large DC deflection," in 3rd *International Conference on Micro Optical Electro Mechanical Systems (Optical MEMS) (MOEMS '99)*, Mainz, Germany, Aug 30-Sept 1, 1999, pp. 239-243.

[43] M. Pai, and N.C. Tien, "Current-controlled bi-directional electrothermally actuated vibromotor," in *Proceedings of the International Conference on Solid-State Sensors and Actuators (Transducers '99)*, June 1999, Sendai, Japan, pp.1764-1767.

[44] J. Buhler, J. Funk, J.G. Korvink, F.-P. Steiner, P.M. Sarro, and H. Baltes, "Electrostatic aluminum micromirrors using double-pass metallization," *Journal of Microelectromechanical Systems*, vol. 6, no. 2, pp. 126-135, June 1997.

[45] H. Toshiyoshi, W. Piyawattanametha, C.-T. Chan, and M.C. Wu, "Linearization of electrostatically actuated surface micromachined 2-D optical scanner," *Journal of Microelectromechanical Systems*, vol. 10, no. 2, pp. 205-214, June 2001.

[46] X.M. Zhang, F.S. Chau, C. Quan, Y.L. Lam, and A.Q. Liu, "A study of the static characteristics of a torsional micromirror," *Sensors and Actuators A (Physical)*, vol. 90, pp. 73-81, 2001.

[47] M.J. Daneman, N.C. Tien, O. Solgaard, A.P. Pisano, K.Y. Lau, and R.S. Muller, "Linear microvibromotor for positioning optical components," *Journal of Microelectromechanical Systems*, vol. 5, no. 3, pp. 159-165, September 1996.

[48] J. Judy, and R.S. Muller, "Magnetic microactuation of torsional polysilicon structures," *Sensors and Actuators A (Physical)*, vol. 53, pp. 392-397, 1996.

[49] W.C. Tang, T.-C. H. Nguyen, and R.T. Howe, "Laterally driven polysilicon resonant microstructures," *Sensors and Actuators A (Physical)*, vol. 20, pp. 25-32, 1989.

[50] A. Friedberger and R. S. Muller, "Improved surface-micromachined hinges for fold-out structures," *Journal of Microelectromechanical Systems*, vol. 7, pp. 315–319, 1998.

[51] E. Smela, O. Ingana, and I. Lundstro, "Controlled folding of micrometer-size structures," *Science*, vol. 268, pp. 1735-1738, 1995.

[52] R.R.A. Syms, C. Gormley, and S. Blackstone, "Improving yield, accuracy and complexity in surface tension self-assembled MOEMS," *Sensors & Actuators A (Physical)*, vol. 88, pp. 273-283, 2001.

[53] R.R.A. Syms, "Surface tension powered self-assembly of 3-D micro-optomechanical structures," *Journal of Microelectromechanical Systems*, vol. 8, no. 4, pp. 448-455, December 1999.

[54] L.Y. Lin, S.S. Lee, K.S.J. Pister, and M.C. Wu, "Micro-machined three-dimensional micro-optics for integrated free-space optical system," *IEEE Photonics Technology Letters*, vol. 6, no. 12, pp. 1445-1447, December 1994.

[55] V.A. Aksyuk, F. Pardo, and D.J. Bishop, "Stress-induced curvature engineering in surface-micromachined devices," in *Proceedings of the SPIE*, vol. 3680, 1999, pp. 984-993.

[56] D.M. Burns and V.M. Bright, "Optical power induced damage to microelectromechanical mirrors," *Sensors and Actuators A (Physical)*, vol. 70, no.1-2, pp.6-14, 1998.

[57] D.M. Tanner, J.A. Walraven, L.W. Irwin, M.T. Dugger, N.F. Smith, W.M. Miller, and S.L. Miller, "The effect of humidity on the reliability of a surface micromachined microengine", in *Proceedings of the IEEE International Reliability Physics Symposium*, 1999, pp. 189-197.

[58] S.B. Brown, W. van Arsdell, and C.L. Muhlstein, "Material reliability in MEMS devices," *Proceedings of International Solid State Sensors And Actuators Conference (Transducers '97)*, 16-19 June, 1997, Chicago, IL, USA, p 591-593.

[59] M.-H. Kiang, O. Solgaard, and K.-Y. Lau, "Electrostatic Combdrive-Actuated Micromirrors for Laser-Beam Scanning and Positioning," *Journal of Microelectromechanical Systems*, vol. 7, no. 1, pp. 27-37, March 1998.

[60] Y. Okada, and Y. Tokumaru, "Precise determination of lattice parameter and thermal expansion coefficient of silicon between 300 and 1500K," *Journal of Applied Physics*, vol. 56, no.2, pp. 314-20, 15 July 1984.

[61] J.-H. Chae, J.-Y. Lee, and S.-W. Kang, "Measurement of thermal expansion coefficient of poly-Si using microgauge sensors," *Sensors and Actuators A (Physical)*, vol. 75, pp. 222-229, June 1999.

[62] Landolt-Bornstein Crystal and Solid State Physics, Ed. K.-H. Hellwege vol..11 (Springer 1979) p.116.

[63] W. Van Arsdell and S. B. Brown, "Subcritical crack growth in silicon MEMS," *Journal of Microelectromechanical Systems*, vol. 8, pp. 319–327, 1999.

[64] ANSI/ISA-S82.01-94, Standards library for measurement and control: guidelines for quality, safety, & productivity, Instrument Society of America, 1993.

[65] S.A McAuley, H. Ashraf, L. Atabo, A. Chambers, S. Hall, J. Hopkins, andG. Nicholls, "Silicon Micromachining using a High Density Plasma Source," *Journal of Physics D*, vol. 34, pp. 2769-2774, 2001.

[66] K. Yamada, and T. Kuriyama, "A novel asymmetric silicon micro-mirror for optical scanning display," in *Proceedings of The 11th Annual International Workshop on Micro Electro Mechanical Systems*, January 25-29, 1998, Heidelberg, Germany, pp. 110-115.

[67] C. Marxer, Patrick Griss, and N.F. de Rooij, "A multichannel variable optical attenuator for power management in fiber optic networks," in *Proceedings of the 10th*

International Conference on Solid-State Sensors and Actuators (Transducers '99), June 1999, Sendai, Japan, pp. 798-799.

[68] H. Toshiyoshi, and H. Fujita, "Electrostatic micro torsion mirrors for an optical switch matrix," *Journal of Microelectromechanical Systems*, vol. 5, no. 4, pp. 231-237, December 1996.

[69] S.S. Lee, L.Y. Lin, and M.C. Wu, "Surface micromachined free-space fiber optic switches", *Electronics Letters*, vol. 31, pp. 1481-1482, 1995.

[70] B. Behin, K.Y. Lau, and R.S. Muller, "Magnetically actuated micromirrors for fiber-optic switching," *Technical Digest of the 1998 Solid-State Sensor and Actuator Workshop*, June 8-11, 1998, Hilton Head Island, SC, pp. 273-276.

[71] P.M. Hagelin, U. Krishnamoorthy, C.M. Arft, J.P. Heritage, and O. Solgaard, "Scalable fiber optic switch using micromachined mirrors," in *Proceedings of the 10th International Conference on Solid-State Sensors and Actuators (Transducers '99)*, June 1999, Sendai, Japan, pp. 496-499.

[72] L.Y. Lin, E.L. Goldstein, and R.W. Tkach, "Free-space micromachined optical switches with submillisecond switching time for large-scale optical crossconnects", *IEEE Photonics Technology Letters*, vol. 10, no. 4, pp. 525-527, 1998.

[73] A.S. Dewa and J.W. Orcutt, "Development of a silicon two-axis micromirror for an optical cross-connect," in *IEEE Solid-State Sensor and Actuator Workshop Technical Digest*, June 2-4, 2000, Hilton Head Island, SC, pp. 93-96.

[74] M.C. Wu, L.-Y. Lin, S.-S. Lee, and K.S.J. Pister, "Micromachined free-space integrated micro-optics," *Sensors and Actuators A (Physical)*, vol. 50, pp. 127-134, 1995.

[75] A. Garnier, T. Bourouina, H. Fujita, T. Jiramoto, E. Orsier, and J.-C. Peuzin, "Contactless actuation of bending and torsional vibrations for 2d-optical-scanner application," in *Proceedings of the 10th International Conference on Solid-State Sensors and Actuators (Transducers '99)*, June 1999, Sendai, Japan, pp. 1876-1877.

[76] J.-H. Lee, Y.-C. Ko, D.-H. Kong, J.-M. Kim, K.B. Lee, and D.-Y. Jeon, "Design and fabrication of scanning mirror for laser display," *Sensors and Actuators A (Physical)*, vol. 96, pp. 223-230, 2002.

[77] C. Strandman, L. Rosengren, H.G.A. Elderstig, and Y. Backlund, "Fabrication of a 45° mirrors together with well-defined V-grooves using wet anisotropic etching of silicon," *Journal of Microelectromechanical Systems*, vol. 4, no. 4, pp 213-219, December 1995.

[78] A.A. Yasseen, J.N. Mitchell, D.A. Smith, and M. Mehregany, "High-aspect-ratio rotary polygon micromotor scanners," *Sensors and Actuators A (Physical)*, vol. 77, pp. 73-79, 1999.

[79] W.-H. Juan, S.W. Pang, "High-Aspect-Ratio Si Vertical Micromirror Arrays for Optical Switching," *Journal of Microelectromechanical Systems*, vol. 7, no. 2, pp. 207-213, June 1998.

[80] N. Asada, M. Takeuchi, V. Vaganov, N. Belov, S. in't Hout, I. Sluchak, "Silicon micro-optical scanner," *Sensors and Actuators A (Physical)*, vol. 83, pp. 284-290, 2000.

[81] V. Milanovic, M. Last, and K.S.J. Pister, "Torsional micromirrors with lateral actuators," in *Proceedings of the 12th International Conference on Solid-State Sensors and Actuators (Transducers '01)*, June 2001, Munich, Germany.

[82] J. Drake and H. Jerman, "A micromachined torsional mirror for track following in magneto-optical disk drives," in *IEEE Solid-State Sensor and Actuator Workshop Technical Digest*, June 2-4, 2000, Hilton Head Island, SC, pp. 10-13.

[83]   A. Garnier, T. Bourouina, H. Fujita, T. Hiramoto, E. Orsier, and J-C. Peuzin, "Contactless actuation of bending and torsional vibrations for 2d-optical-scanner application," *Proceedings of the 10th International Conference on Solid-State Sensors and Actuators (Transducers '99)*, 1999, Sendai, Japan, pp. 1876-1877.

[84]   Z.J. Yao, and N.C. MacDonald, "Single crystal silicon supported thin film micromirrors for optical applications," *Optical Engineering*, vol. 36 , no.5, pp. 1408-13, 1997.

[85]   J.-L.A Yeh, J. Hongrui, and N.C. Tien, "Integrated polysilicon and DRIE bulk silicon micromachining for an electrostatic torsional actuator.", *Journal of Microelectromechanical Systems*, vol. 8, no.4, pp. 456-65, 1999.

[86]   W.P. Maszara, G. Goetz, A. Caviglia, and J.B. McKitterick, "Bonding of silicon wafers for silicon-on-insulator," *Journal of Applied Physics*, vol. 64, no. 10, pp. 4943-4950, 15 November 1988.

[87]   G.R. Elion, H.A. Elion, Electro-Optics Handbook, New York: Marcel Dekker, Inc., 1979.

[88]   M.-H. Kiang, O. Solgaard, R.S. Muller, and K.-Y. Lau, "Micromachined polysilicon microscanners for barcode readers," *IEEE Photonics Technology. Letters*, vol. 8 pp. 95-97, 1998.

[89]   A. Friedberger, and R.S. Muller, "Improved surface-micromachined hinges for fold-out structures," *Journal of Microelectromechanical Systems*, vol. 7, pp. 315-319, 1998.

[90]   H. Schenk, P. Durr, D. Kunze, H. Lakner, and H. Kuck, "An electrostatically excited 2D-micro-scanning-mirror with an in-plane configuration of the driving electrodes," *Proc. IEEE 13th Annual International Conference on Micro Electro Mechanical Systems*, 2000, Miyazaki, Japan, pp. 473-478.

[91]   H. Schenk, P. Durr, D. Kunze, H. Lakner, and H. Kuck, "A resonantly excited 2D-micro-scanning-mirror with large deflection," *Sensors and Actuators A (Physical)*, vol. 89, pp. 104-111, 2001.

[92]   J.-L. A. Yeh, Y.-Y. Hui, and N.C. Tien, "Electrostatic model for an asymmetric combdrive," *Journal of Microelectromechanical Systems*, vol. 9 no. 1, pp. 126-135, March 2000.

[93]   L. J. Hornbeck, "Digital Light Processing™ for High-Brightness, High-Resolution Applications," *Electronic Imaging, EI '97*, Projection Displays III. San Jose, CA, February 10-12, 1997.

[94]   O. Solgaard, S.F.A. Sandejas, and D.M. Bloom, "Deformable Grating Optical Modulator," *Optics Letters*, vol. 17, pp. 688-690, 1992.

[95]   D.T. Amm and R.W. Corrigan, "Optical performance of the grating light valve technology," *Proceedings of the SPIE – The International Society for Optical Engineering*, 1999, vol. 3634, 1999, pp. 71-78.

[96]   D. A. Francis, M.-H. Kiang, O. Solgaard, and K.-Y. Lau, "Compact 2D laser beam scanner with fan laser array and Si micromachined microscanner." *Electronics Letters*, vol.33, no.13, pp. 1143-5, 19 June 1997.

[97]   P.M. Hagelin and O. Solgaard, "Optical raster-scanning displays based on surface-micromachined polysilicon mirrors". *IEEE Journal of Selected Topics in Quantum Electronics*, vol. 5, no. 1, pp. 67-74, Jan.-Feb. 1999.

[98]   R. H. Webb, "Confocal scanning laser opthalmoscope," *Applied Optics*, vol. 26, no. 8, pp. 1492-9, 15 April 1987.

[99] Y. Hwang, J. Park, Y. Kim, and J. Kim, "Large-scale full color laser projection display," in *Proceedings of the 18th International Display Research Conference ASIA Display '98*, Sept. 28-Oct. 1, 1998, pp. 1167-1170.

[100] K. Yamada and T. Kuriyama, "A novel asymmetric silicon micro-mirror for optical scanning display," in *Proceedings of the 11th Annual International Workshop on Micro Electro Mechanical Systems*, January 25-29, 1998, Heidelberg, Germany, pp. 110-115.

[101] P. Hagelin, K. Cornett, and O. Solgaard: "Micromachined mirrors in a raster scanning display system," in *Proceedings of the 1998 IEEE/LEOS Summer Topical Meetings: Optical MEMS*, 20-24 July, 1998, Monterey, CA, pp. 109-110.

[102] J. Hagerman, "Optimum spot size for raster-scanned monochrome CRT displays", *Journal of the SID*, vol. 1, no. 3, pp. 367-369, 1993